SO-AHP-954

# ON THE JOB SERIES

# REAL PEOPLE WORKING *in*

# MECHANICS, INSTALLATION, AND REPAIR

Blythe Camenson

## VGM Career Horizons

*NTC/Contemporary Publishing Group*

**Library of Congress Cataloging-in-Publication Data**

Camenson, Blythe.
    On the job : Real people working in mechanics, installation, and repair /
Blythe Camenson.
        p.    cm. — (On the job series)
      ISBN 0-658-00017-9 (cloth). — ISBN 0-658-00018-7 (pbk.)
      1. Machinery—Maintenance and repair Vocational guidance.
    2. Mechanics (Persons) Vocational guidance.    I. Title.    II. Title:
Mechanics, installation, and repair.    III. Series.
    TJ157.C36    1999
    621.8′023—dc21                                          99-34060
                                                            CIP

Cover design by Nick Panos

Published by VGM Career Horizons
A division of NTC/Contemporary Publishing Group, Inc.
4255 West Touhy Avenue, Lincolnwood (Chicago), Illinois 60712-1975 U.S.A.
Copyright © 2000 by NTC/Contemporary Publishing Group, Inc.
All rights reserved. No part of this book may be reproduced, stored in a retrieval
system, or transmitted in any form or by any means, electronic, mechanical,
photocopying, recording, or otherwise, without the prior written permission of
NTC/Contemporary Publishing Group, Inc.
Printed in the United States of America
International Standard Book Number: 0-658-00017-9 (cloth)
                                    0-658-00018-7 (paper)
00  01  02  03  04  05  VL  18  17  16  15  14  13  12  11  10  9  8  7  6  5  4  3  2  1

# REAL PEOPLE WORKING *in*

# MECHANICS, INSTALLATION, AND REPAIR

YA
331.702
M464C

# Contents

# Acknowledgments

The author would like to thank the following people for providing information about their careers.

- J. G. Barr, automotive mechanic

- Pete Bennett, elevator mechanic

- Stan Clay, aircraft maintenance supervisor

- Robert Cravey, Jr., airframe and powerplant mechanic

- Adam Dodds, tractor mechanic

- Jim Foreman, musical instrument repairer

- Harry Forman, automotive mechanic

- Juan Garcia, heavy equipment mechanic

- Thérèse Heckel, switching equipment technician

- Robert Holland, owner of a vending machine sales and repair business

- Michael John, computer bench technician

- Troy Scott McClure, owner of a refrigeration, heating, and air conditioning business

- Roger Pryor, transmission builder

- Joe Robertson, small engine mechanic

- Michael Sanchez, owner of a power tool sales and repair business

- Thomas Walker, field maintenance worker

- Kirk Woodruff, heavy equipment mechanic

# How to Use This Book

*On the Job: Real People Working in Mechanics, Installation, and Repair* is part of a series of career books designed to help you find essential information quickly and easily. Unlike other career resources on the market, books in the *On the Job* series provide you with information on hundreds of careers, in an easy-to-use format. This includes information on:

- the nature of the work

- working conditions

- employment

- training, other qualifications, and advancement

- job outlooks

- earnings

- related occupations

- sources of additional information

But that's not all. You'll also benefit from a first-hand look at what jobs are really like, as told in the words of the employees themselves. Throughout the book, one-on-one interviews with dozens of practicing professionals enrich the text and enhance your understanding of life on the job.

These interviews tell what each job entails, what the duties are, what the lifestyle is like, and what the upsides and downsides are. All of the professionals reveal what drew them to the field and how they got started. And to help you make the best career choice for yourself, each professional offers you some expert advice based on years of personal experience.

Each chapter also lets you see at a glance, with easy-to-use reference symbols, the level of education required and salary range for the featured occupations.

So, how do you use this book? Easy. You don't need to run to the library and bury yourself in cumbersome documents from the Bureau of Labor Statistics, nor do you need to rush out and buy a lot of bulky books you'll never plow through. All you have to do is glance through our extensive table of contents, find the fields that interest you, and read what the experts have to say.

# Introduction to the Field

If you like to work with your hands and are good at it, chances are you're already considering a career in which you can put your talents to good use. Jobs in mechanics, repair, and installation are plentiful and can allow you to make a good living doing something you love.

Glancing through this book's table of contents will give you an idea of all the choices open to you in these fields. But perhaps you're not sure of the working conditions the different areas offer or which area would best suit your personality, skills, and lifestyle. There are several factors to consider when deciding which sector to pursue. Each field carries with it different skill levels and responsibility. To identify occupations that will match your expectations, you need to know what each job entails.

Ask yourself the following questions and make note of your answers. Then, as you go through the following chapters, compare your requirements to the information provided by the professionals interviewed inside. Their comments will help you pinpoint the areas that would interest you and eliminate those that would clearly be the wrong choice.

- How much time are you willing to commit to training? Some skills can be learned on the job; others can take much longer and require a year or more of formal training.

- How much of a people person are you? Some jobs, such as aircraft mechanic, offer few opportunities for personal contact; other positions, such as vending machine repairer, require significant interaction with others.

- How much time away from home are you willing to spend? Some jobs will have you working odd and long hours, while others will give you free weekends.

- How much money do you expect to earn as a beginner, and then as an employee with a few years' experience? In general, those areas that pay the most also require the largest investment of time for training.

- Will the work offer enough challenge? Will it provide you with a sense of accomplishment, or will it become tedious after you've learned the ropes?

- How much independence do you require? Do you want to be your own boss, or will you be content as a salaried employee?

- How much stress can you handle? Would you prefer to avoid work that could be emotionally draining?

Knowing what your expectations are and comparing them to the realities of particular jobs will help you make informed choices.

Although *On the Job: Real People Working in Mechanics, Installation, and Repair* strives to be as comprehensive as possible, not all jobs in this extensive field have been covered or given the same amount of emphasis. If you still have questions after reading this book, there are a number of other avenues to pursue. You can get more information by contacting the sources listed at the end of each chapter. You can also find professionals on your own to talk to and observe as they go about their work. Any remaining gaps you discover can be filled by referring to the *Occupational Outlook Handbook*, available in most libraries online at http://stats.bls.gov/ocohome.htm.

ON THE
**JOB**
SERIES

# REAL PEOPLE
# WORKING *in*

# MECHANICS,
# INSTALLATION,
# AND REPAIR

# CHAPTER 1 Aircraft Mechanics

**EDUCATION**
H.S. Preferred
License (Other)

**$$$ SALARY**
$23,000 to $48,000

## OVERVIEW

To keep aircraft in peak operating condition, aircraft mechanics and engine specialists perform scheduled maintenance, make repairs, and complete inspections required by the Federal Aviation Administration (FAA).

Many aircraft mechanics specialize in preventive maintenance. Following a schedule based on the number of hours the aircraft has flown, calendar days, cycles of operation, or a combination of these factors, mechanics inspect the engines, landing gear, instruments, pressurized sections, accessories (brakes, valves, pumps, and air conditioning systems, for example), and other parts of the aircraft and do the necessary maintenance. They may examine an engine through specially designed openings while working from ladders or scaffolds, or use hoists or lifts to remove the entire engine from the craft. After taking the engine apart, mechanics may use precision instruments to measure parts for wear, and use X-ray and magnetic inspection equipment to check for invisible cracks. Worn or defective parts are repaired or replaced. They also may repair sheet metal or composite surfaces, measure the tension of control cables, or check for corrosion, distortion, and cracks in the fuselage, wings, and tail. After completing all repairs, mechanics must test the equipment to ensure that it works properly.

Mechanics specializing in repair work rely on the pilot's description of a problem to find and fix faulty equipment. For example, during a preflight check, a pilot may discover that the aircraft's fuel gauge does not work. To solve the problem, mechanics may check the electrical connections, replace the gauge, or use electrical test equipment to make sure no wires are broken or shorted out. They work as fast as safety permits so that the aircraft can be put back into service quickly.

Mechanics may work on one type or many different types of aircraft, such as jets, propeller-driven airplanes, and helicopters; or, for efficiency, they may specialize in one section of a particular type of aircraft, such as the engine, hydraulic, or electrical system. As a result of technological advances, mechanics spend an increasing amount of time repairing electronic systems such as computerized controls. They also may be required to analyze and develop solutions to complex electronic problems. In small, independent repair shops, mechanics usually inspect and repair many different types of aircraft.

Mechanics ordinarily work in hangars or in other indoor areas, though they may work outdoors—sometimes in unpleasant weather—when the hangars are full or when repairs must be made quickly. This occurs most often to airline mechanics who work at airports, because minor repairs and preflight checks might be made at the terminal to save time. Mechanics often work under time pressure to maintain flight schedules or, in general aviation, to keep from inconveniencing customers. At the same time, mechanics have a tremendous responsibility to maintain safety standards, and this can make the job stressful.

Frequently, mechanics must lift or pull objects weighing as much as seventy pounds. They often stand, lie, or kneel in awkward positions and occasionally must work in precarious positions on scaffolds or ladders. Noise and vibration are common when testing engines. Aircraft mechanics generally work forty hours a week on eight-hour shifts around the clock. Overtime work is frequent.

# TRAINING

The majority of mechanics who work on civilian aircraft are certified by the FAA as *airframe mechanic, powerplant mechanic,* or *repairer.* Airframe mechanics are authorized to work on any part of the aircraft except the instruments, powerplants, and propellers. Powerplant mechanics are authorized to work on engines and to do limited work on propellers. Repairers, who are employed by FAA-certified repair stations and air carriers, work on instruments and propellers.

Combination airframe-and-powerplant mechanics—called A&P mechanics—can work on any part of the plane, and those with an inspector's authorization can certify inspection work completed by other mechanics. Uncertified mechanics are supervised by those with certificates.

The FAA requires at least eighteen months of work experience for an airframe, powerplant, or repairer's certificate. For a combined A&P certificate, at least thirty months of experience working with both engines and airframes are required. To obtain an inspector's authorization, a mechanic must have held an A&P certificate for at least three years. Applicants for all certificates also must pass written and oral tests and demonstrate that they can do the work authorized by the certificate. Most airlines require that mechanics have a high school diploma and an A&P certificate.

Although a few people become mechanics through on-the-job training, most learn their job in one of about 192 trade schools certified by the FAA. Student enrollment in these schools varies greatly; some have as few as 50 students while at least one school has about 800 students. FAA standards established by law require that certified mechanic schools offer students a minimum of 1,900 actual class hours. Courses in these trade schools generally last from twenty-four to thirty months and provide training with the tools and equipment used on the job. For an FAA certificate, attendance at such schools may substitute for work experience. However, these schools do not guarantee jobs or FAA certificates. Aircraft trade schools are placing more emphasis on newer technologies such as turbine engines, aviation electronics, and composite materials—including graphite, fiberglass, and boron—all of which are increasingly being used in the construction of new aircraft. Less emphasis is being placed on older technologies such as woodworking and welding. Employers prefer mechanics who can perform a wide variety of tasks. Mechanics learn many different skills in their training that can be applied to other jobs.

Some aircraft mechanics in the armed forces acquire enough general experience to satisfy the work experience requirements for the FAA certificate. With additional study, they may pass the certifying exam. Generally, however, jobs in the military are too specialized to provide the broad experience required by the FAA. Most mechanics have to complete the entire training program at a trade school, although a few receive some credit for the material they learned in the service. In any case, military experience is a great advantage when seeking employment; employers consider trade school graduates who have this experience to be the most desirable applicants.

Courses in mathematics, physics, chemistry, electronics, computer science, and mechanical drawing are helpful because many of their principles are involved in the operation of an aircraft, and knowledge of the principles often is necessary to make repairs. Courses that develop writing skills are also important because mechanics are often required to submit reports.

As new and more complex aircraft are designed, more employers are requiring mechanics to take ongoing training to update their skills. Recent technological advances in aircraft maintenance necessitate a strong background in electronics for both acquiring and retaining jobs in this field. New FAA certification standards will make ongoing training mandatory. Every twenty-four months, mechanics will be required to take at least sixteen hours of training to keep their certificate. Many mechanics take courses offered by manufacturers or employers, usually through outside contractors.

Aircraft mechanics must do careful and thorough work that requires a high degree of mechanical aptitude. Employers seek applicants who are self-motivated, hard-working, enthusiastic, and able to diagnose and solve complex mechanical problems. Agility is important for the reaching and climbing necessary for the job. Because they may work on the top of wings and fuselages on large jet planes, aircraft mechanics must not be afraid of heights.

As aircraft mechanics gain experience, they have the opportunity for advancement. Opportunities are best for those who have an aircraft inspector's authorization. A mechanic may advance to lead mechanic (or crew chief), inspector, lead inspector, or shop supervisor. In the airlines, where promotion is often determined by examination, supervisors may advance to executive positions. Those with broad experience in maintenance and overhaul have become inspectors with the FAA. With additional business and management training, some open their own aircraft maintenance facilities.

# JOB OUTLOOK

Aircraft mechanics held about 137,000 jobs in 1996. Over three-fifths of all salaried mechanics worked for airlines, nearly one-fifth for aircraft assembly firms, and nearly one-sixth for the federal government. Most of the rest were general aviation mechanics, the majority of whom worked for independent repair shops or companies that operate their own planes to transport executives and cargo. Very few mechanics were self-employed.

Most airline mechanics work at major airports near large cities. Civilian mechanics employed by the armed forces work at military installations. A large proportion of mechanics who work for aircraft assembly firms are located in California or Washington. Others work for the FAA, many at the facilities in Oklahoma City, Atlantic City, or Washington, D.C. Mechanics for independent repair shops work at airports in every part of the country.

There should be an improved outlook for aircraft mechanics over the next ten years. The smaller numbers of younger workers in the labor force, together with fewer entrants from the military and a larger number of retirements, should mean more favorable employment conditions for students just beginning training.

Job prospects for aircraft mechanics are expected to vary among types of employers. Opportunities are likely to be the best at the smaller commuter and regional airlines, at FAA repair stations, and in general aviation. Because wages in these companies tend to be relatively low, there are fewer applicants for these jobs than for jobs with the major airlines. Also, some jobs will become available as experienced mechanics leave for higher-paying jobs with airlines or transfer to another occupation. Mechanics will face more competition for airline jobs because the high wages and travel benefits attract more qualified applicants than there are openings. Prospects will be best for applicants with significant experience. Mechanics who keep abreast of technological advances in electronics, composite materials, and other areas will be in greatest demand. The number of job openings for aircraft mechanics in the federal government should decline as the size of the armed forces is reduced.

Employment of aircraft mechanics is expected to increase about as rapidly as the average for all occupations through 2006. A growing population and rising incomes are expected to stimulate the demand for airline transportation, and the number of aircraft is expected to increase as well. However, employment growth will be restricted somewhat by increases in productivity resulting from greater use of automated inventory control and modular systems that speed repairs and parts replacement.

Most job openings for aircraft mechanics through 2006 will stem from replacement needs. Each year, as mechanics transfer to other occupations or retire, several thousand job openings will arise. Aircraft mechanics have a comparatively strong attachment to the occupation, reflecting their significant investment in training and a love for aviation. However, because aircraft mechanics' skills are transferable to other occupations, some mechanics leave for work in related fields.

Declines in air travel during recessions force airlines to curtail the number of flights, which results in less aircraft maintenance and, consequently, layoffs for aircraft mechanics.

## SALARIES

In 1996, the median annual salary of aircraft mechanics was about $35,000. The middle 50 percent earned between $29,000 and $44,000. The top 10 percent of all aircraft mechanics earned over $48,000 a year, and the bottom 10 percent earned less than $23,200. Mechanics who worked on jets generally earned more than those working on other aircraft. Airline mechanics and their immediate families receive reduced-fare transportation on their own and most other airlines.

Earnings of airline mechanics are generally higher than mechanics working for other employers. Average hourly pay for beginning aircraft mechanics was estimated to range from $18 at the smaller turbo-prop airlines to $22 at the major airlines in 1996. Earnings of experienced mechanics were estimated to range from $25 to $32 an hour.

Almost one-half of all aircraft mechanics, including those employed by some major airlines, are covered by union agreements. The principal unions are the International Association of Machinists and Aerospace Workers and the Transport Workers Union of America. Some mechanics are represented by the International Brotherhood of Teamsters.

## RELATED FIELDS

Workers in some other occupations that involve similar mechanical and electrical work are electricians, elevator repairers, and telephone maintenance mechanics.

## INTERVIEW
### Stan Clay
### Aircraft Maintenance Supervisor

Stan Clay is an aircraft maintenance supervisor with Airborne Express in Wilmington, Ohio. He's been in aviation for more than twenty-six years and in large transport aircraft maintenance for fourteen years. He attended airframe and powerplant mechanics school at the Spartan School of Aeronautics in Tulsa, Oklahoma, from 1973 to 1975. He earned his bachelor of science in air commerce with a minor in flight technology from the Florida Institute of Technology in Melbourne in 1984. He is currently working on his masters of aviation science at Embry-Riddle Aeronautical University in Cincinnati, Ohio.

## How Stan Clay Got Started

"My father, who had forty-two years with Eastern Airlines, was a very well-respected general foreman in the maintenance department at Eastern. He had an unbelievable ability to make anything with his hands that he could see in his mind. As a result, he was in great demand, not only by Eastern but also by other airlines and the aircraft manufacturers themselves. He designed and made tools that are now used virtually all over the industry. As I grew up, I watched him and tried to emulate his mechanical abilities. Airplanes fascinated me, and I earned my pilot's license before I graduated from high school. Because I had grown up in a mechanical atmosphere and was taken with aviation, it just seemed natural to pursue a career in this field.

"After high school, I went to Spartan School of Aeronautics in Tulsa, Oklahoma. The school specializes in aviation maintenance training. Upon graduation, students take the FAA test for their airframe and powerplant mechanics license.

"Immediately after graduation from Spartan, I joined the United States Army as a helicopter repairman. When I got out of the army, I went to college and was subsequently hired by Eastern Airlines. (My father had long since retired.)

"I advanced quickly through the ranks to become the manager of aircraft structures and technical support and was enjoying a very rewarding career when Eastern ran into trouble with its unions. I was the last structures maintenance person to leave Eastern's main base in Miami, Florida.

"Like everyone else, I had to scramble for a job. I sent out tons of resumes throughout the aviation community. Of the job offers that came back, Airborne Express seemed the most promising. I interviewed and was offered the position of aircraft structures supervisor—which is one of two supervisory positions on any one shift. About five years ago, the other supervisor on my shift retired. I was given the assignment to cover both positions until his replacement could be hired. The company is no longer looking for a replacement. I have become the sole supervisor for not only the structures department but also for the maintenance department."

## What the Job's Really Like

"This job runs the gamut. I get involved in research as well as hands-on work on the aircraft. Some things are routine and some are totally new.

"I have a crew of eleven mechanics and twelve structures technicians. (The difference between the two is that a mechanic deals with components such as engines, hydraulic systems, or flight control systems. A structures technician deals with the actual airframe components such as wings, fuselage, and tail.)

"My workday begins by checking E-mail, which, among other things, contains reports from other shifts and departments on anything done on our particular aircraft during the preceding twenty-four hours. Next is a meeting with the other department heads in which the game plan for the next twenty-four hours is mapped out. From there, individual assignments are given to the workforce. This begins my favorite part of the day. Inspectors have recorded discrepancies they have found on our aircraft, and they report those discrepancies to us for repair. Each one of those items will be evaluated, and a specific repair will be suggested. Most are relatively easy to deal with, but some are extremely involved and may require thousands of man-hours. Since no two discrepancies are exactly alike, it takes creativity to come up with repairs that will restore the structural member or component to its original strength.

"During the course of an aircraft check, I am in contact with the manufacturer (Douglas or Boeing, for example) at least once. Everything is held to very exacting standards to ensure the safety and reliability of the aircraft.

"Of course, every check is done on a time schedule. Things can get pretty intense when it comes down to the end of a check. I love it! The mark of your success is the record of maintenance reliability that the aircraft has following your check, as well as whether it came out of the check on time. If you can succeed at both consistently (as our crew does), you are in the minority.

"I like being able to encounter and fix virtually any problem within the scheduled time of the check. However, this means nothing if the aircraft is not safe and reliable. You must have both. Achieving a long track record of both safety and reliability is what I like the best.

"The thing I like the least is today's 'anything goes' mind-set with respect to workforce. I believe that companies should provide for their employees, but I need people who will show up when they are supposed to and give a day's work for a day's pay. We have too many requirements for hiring or promoting this or that individual and too much legislation like the Family and Medical Leave Act, which is burdensome to employers. I don't like outside involvement when none has been warranted. Speaking frankly to an employee about his or her performance today is all but impossible."

## Expert Advice

"Anyone wanting to get into aviation maintenance has got to start with a license. There are plenty of trade schools and universities that can provide the necessary training.

"The next step is to get your foot in the door somewhere. You can do this at the same time as you are getting your education. I worked at a small airport that was affiliated with the university while I attended college. You could load baggage for an airline and, after obtaining your license, apply for a transfer to the maintenance department. Many airlines will go inside to fill job openings prior to looking outside the company.

"The military is certainly not a bad idea. They will take you with no experience and then pay for your training.

"I make in the neighborhood of $60,000 a year. A mechanic just starting out can expect to make around $13 an hour. Progression is pretty quick, and rates can go as high as $20 to $25 an hour."

## INTERVIEW
### Robert Cravey, Jr.
## Airframe and Powerplant Mechanic

Robert Cravey, Jr., works as an airframe and powerplant mechanic for a privately owned company that does corporate and government aircraft maintenance. He received his A&P license in 1972.

## How Robert Cravey, Jr., Got Started

"A visit to Ellington Air Force Base in Texas with my family inspired me to choose this career. I attended U.S. Air Force schools, civil college courses, and manufacturers' schools on military and civil aircraft and helicopters to obtain formal training. I am also a commercial pilot with multiengine and instrument ratings and 2,000-plus hours of flight time.

"For my first job, I found a small operator at an airport in Texas and set a resume before him. He needed a mechanic, I needed a job. It worked—for a while. Then he went bankrupt.

"For my current job, I applied to the home office in Nebraska and interviewed twice for the only position available. Five candidates were interviewed. I was the only successful one."

## What the Job's Really Like

"My job is to ensure that the aircraft I do maintenance on are airworthy. My job involves maintaining two Shorts 360 'Sherpa' (called C-23/B+) aircraft for the army national guard. It is a nationwide contract program. The aircraft are on twenty-four-hour call and can go anyplace the army needs them. It is a relaxed atmosphere, but demanding in terms of hours and travel. Anytime, anyplace, and all hours. No excuses.

"My hours can be from 40 to 100 per week. Typical day? No such animal . . . in at 8:30 A.M. and out when the airplane is ready to fly. Or if parts are on order, I come in again when they arrive. Then I fix the airplane, leave, return to launch missions, and then take care of support equipment, paperwork, shipping, receiving, faxes, and telephone calls. I travel with the airplane if need be. I'm on call every other weekend . . . no boring moments here!

"I have a great lead man to work for. There are only two of us at the site presently. He's down-to-earth and honest. There are no downsides to my job—none."

### Expert Advice

"You shouldn't expect any glamour or high salaries. Unless you are born into the airline world and inherit tools, be prepared to spend a lot for tools.

"You also need to be prepared to go anywhere to work. You must be capable of precision, honesty, and writing legibly.

"Be flexible in your attitude toward employers, and most of all treat the customer with the utmost respect—they pay the wage!

"Never believe there is a person who knows it all. There is always room to learn. You should obtain all the manufacturers' schooling you can, obtain the I.A., and never stop being curious about airplanes."

# FOR MORE INFORMATION

Information about jobs in a particular airline may be obtained by writing to the personnel manager of the company. For general information about aircraft mechanics, write to:

Professional Aviation Maintenance Association
1200 18th Street NW, Suite 401
Washington, DC 20036

For information on jobs in a particular area, contact employers at local airports or at local offices of the state employment service.

# CHAPTER 2 Automotive Mechanics

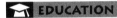
**EDUCATION**
H.S. Preferred

**$$$ SALARY**
$13,000 to $45,000

## OVERVIEW

Automotive mechanics, often called *automotive service technicians*, repair and service automobiles and occasionally light trucks, such as vans and pickups, with gasoline engines. (Motorcycle mechanics—who repair and service motorcycles, motorscooters, mopeds, and occasionally small all-terrain vehicles—are discussed in Chapter 3.)

Anyone whose car or light truck has broken down knows the importance of the mechanic's job. The ability to diagnose the source of the problem quickly and accurately, one of the mechanic's most valuable skills, requires good reasoning ability and a thorough knowledge of automobiles. In fact, many mechanics consider diagnosing hard-to-find troubles one of their most challenging and satisfying duties.

When mechanical or electrical problems occur, mechanics first get a description of the symptoms from the owner or, if they work in a dealership or large shop, the repair service estimator who wrote the repair order. The mechanic may have to test drive the vehicle or use a variety of testing equipment, such as engine analyzers, spark plug testers, or compression gauges, to locate the problem. Once the cause of the problem is found, mechanics make adjustments or repairs. If a part is damaged or worn beyond repair, or cannot be fixed at a reasonable cost, it is replaced, usually after consultation with the vehicle owner.

During routine service, mechanics inspect, lubricate, and adjust engines and other components, repairing or replacing parts before they cause breakdowns. They usually follow a checklist to be sure they examine all important parts, such as belts, hoses, steering systems, spark plugs, brake and fuel systems, wheel bearings, and other potentially troublesome items.

Mechanics use a variety of tools in their work. They use power tools such as pneumatic wrenches to remove bolts quickly; machine tools such as lathes and grinding machines to rebuild brakes and other parts; welding and flame-cutting equipment to remove and repair exhaust systems and other parts; jacks and hoists to lift cars and engines; and a growing variety of electronic service equipment, such as infrared engine analyzers and computerized diagnostic devices. They also use many common hand tools such as screwdrivers, pliers, and wrenches to work on small parts and get at hard-to-reach places.

Automotive mechanics in larger shops have increasingly specialized their skills. For example, automatic transmission mechanics work on gear trains, couplings, hydraulic pumps, and other parts of automatic transmissions. Because these are complex mechanisms and include electronic parts, their repair requires considerable experience and training, including a knowledge of hydraulics. Tune-up mechanics adjust the ignition timing and valves, and adjust or replace spark plugs and other parts to ensure efficient engine performance. They often use electronic test equipment to help them adjust and locate malfunctions in fuel, ignition, and emissions control systems.

*Automotive air conditioning mechanics* install and repair air conditioners and service components such as compressors and condensers. *Front-end mechanics* align and balance wheels and repair steering mechanisms and suspension systems. They frequently use special alignment equipment and wheel-balancing machines. *Brake repairers* adjust brakes, replace brake linings and pads, repair hydraulic cylinders, turn discs and drums, and make other repairs on brake systems. Some mechanics specialize in both brake and front-end work. *Automotive radiator mechanics* clean radiators with caustic solutions, locate and solder leaks, and install new radiator cores or complete replacement radiators. They also may repair heaters and air conditioners, and solder leaks in gasoline tanks.

The majority of automotive mechanics work for retail and wholesale automotive dealers, independent automotive repair shops, and gasoline service stations.

Most automotive mechanics work a standard forty-hour week, but some self-employed mechanics work longer hours. Generally, mechanics work indoors. Most repair shops are well ventilated and well lit, but some are drafty and noisy. Mechanics frequently work with dirty and greasy parts, and in awkward positions. They often must lift heavy parts and tools. Minor cuts, burns, and bruises are common, but serious accidents are avoided when the shop is kept clean and orderly and safety practices are observed.

# TRAINING

Automotive technology is rapidly increasing in sophistication, and most training authorities strongly recommend that persons seeking trainee automotive mechanic jobs complete a formal training program after graduating from high school. However, some automotive mechanics still learn the trade solely by assisting and working with experienced mechanics.

Automotive mechanic training programs are offered in high schools, community colleges, and public and private vocational and technical schools, but postsecondary programs generally provide more thorough career preparation than high school programs. High school programs, while an asset, vary greatly in quality. Some offer only an introduction to automotive technology and service for the future consumer or hobbyist, while others aim to equip graduates with enough skills to get a job as a mechanic's helper or trainee mechanic after graduation.

Postsecondary automotive mechanic training programs vary greatly in format, but generally provide intensive career preparation through a combination of classroom instruction and hands-on practice. Some trade and technical school programs provide concentrated training for six months to a year, depending on how many hours the student must attend each week. Community college programs normally spread the training out over two years; supplement the automotive training with instruction in English, basic mathematics, computers, and other subjects; and award an associate degree.

The various automobile manufacturers and their participating dealers sponsor two-year associate degree programs at more than 100 community colleges across the nation. The manufacturers provide service equipment and late-model cars on which students practice new skills, and ensure that the programs teach the latest automotive technology. Curricula are updated frequently to reflect changing technology and equipment. Students in these programs typically spend alternate six- to twelve-week periods attending classes full time and working full time in the service departments of sponsoring dealers. Because students spend time gaining valuable work experience, these programs may take as long as four years to complete, instead of the normal two years required to earn an associate degree in automotive service technology. However, they offer students the opportunity to earn money while going to school and promise a job upon graduation. Also, some sponsoring dealers provide students with financial assistance for tuition or the purchase of tools.

The National Automotive Technicians Education Foundation (NATEF), an affiliate of the National Institute for Automotive Service Excellence (ASE), certifies automobile mechanic training programs offered by high schools and postsecondary trade schools, technical institutes, and community colleges. While NATEF certification is voluntary and many institutions have not sought it, certification does signify that the program meets uniform standards for instructional facilities, equipment, staff credentials, and curriculum. More than 850 high school and postsecondary automotive mechanic training programs are certified by NATEF.

Knowledge of electronics is increasingly desirable for automotive mechanics because electronics is being used in a growing variety of automotive components. Engine controls and dashboard instruments were among the first components to use electronics, but now electronics is being used in brakes, transmissions, steering systems, and a variety of other components. In the past, problems involving electrical systems or electronics were usually handled by a specialist, but electronics is becoming so commonplace that most automotive mechanics must be familiar with at least the basic principles of electronics in order to recognize when an electronic malfunction may be responsible for a problem. In addition, automotive mechanics frequently must be able to test and replace electronic components.

For trainee mechanic jobs, employers look for people with good reading and basic mathematics and computer skills who can study technical manuals to keep abreast of new technology. People who have a desire to learn new service and repair procedures and specifications are excellent candidates for trainee mechanic jobs. Trainees also must possess mechanical aptitude and knowledge of how automobiles work. Most employers regard the successful completion of a vocational training program in automotive mechanics at a postsecondary institution as the best preparation for trainee positions. Experience working on motor vehicles in the armed forces or as a hobby is also valuable. Completion of high school is required by a growing number of employers. Courses in automotive repair, electronics, physics, chemistry, English, computers, and mathematics can provide a good basic educational background for a career as an automotive mechanic.

Beginners usually start as trainee mechanics, helpers, lubrication workers, or gasoline service station attendants, and gradually acquire and practice their skills by working with experienced mechanics. Although a beginner can perform many routine service tasks and make simple repairs after a few months' experience, it usually takes one to two years of experience to acquire adequate proficiency to become a journey service mechanic and quickly perform the more difficult types of routine service and repairs. However, graduates of the better postsecondary mechanic training programs are often able to earn promotion to the journey level after only a few months on the job. An additional one to two years are usually required to become thoroughly experienced and familiar with all types of repairs. Difficult specialties, such as transmission repair, require yet another year or two of training and experience.

In contrast, automotive radiator mechanics and brake specialists, who do not need an all-around knowledge of automotive repair, may learn their jobs in considerably less time.

In the past, many people have become automotive mechanics through three- or four-year formal apprenticeship programs. However, as formal automotive training programs have increased in popularity, the number of employers willing to make such a long-term apprenticeship commitment has greatly declined.

Mechanics usually buy their hand tools, and beginners are expected to accumulate tools as they gain experience. Many experienced mechanics have thousands of dollars invested in tools. Employers furnish power tools, engine analyzers, and other test equipment.

Employers increasingly send experienced automotive mechanics to manufacturer training centers to learn to repair new models or to receive special training in the repair of components such as electronic fuel injection or air conditioners. Motor vehicle dealers may also send promising beginners to manufacturer-sponsored mechanic training programs. Factory representatives come to many shops to conduct short training sessions.

Voluntary certification by ASE is widely recognized as a standard of achievement for automotive mechanics. Mechanics are certified in one or more of eight different service areas, such as electrical systems, engine repair, brake systems, suspension and steering, and heating and air conditioning. Master automotive mechanics are certified in all eight areas. For certification in each area, mechanics must have at least two years of experience and pass a written examination; completion of an automotive mechanic program in high school, vocational or trade school, or community or junior college may be substituted for one year of experience. Certified mechanics must retake the examination at least every five years.

Experienced mechanics who have leadership ability may advance to shop supervisor or service manager. Mechanics who work well with customers may become automotive repair service estimators. Some with sufficient funds open independent repair shops.

# JOB OUTLOOK

Automotive mechanics hold about 775,000 jobs in North America. The majority work for retail and wholesale automotive dealers and independent automotive repair shops, and gasoline service stations. Others are employed in automotive service facilities at department, automotive, and home supply stores. A small number maintain automobile fleets for taxicab and automobile leasing companies; federal, state, and local governments; and other organizations.

Motor vehicle manufacturers employ some mechanics to test, adjust, and repair cars at the end of assembly lines.

Approximately 20 percent of automotive mechanics are self-employed.

Job opportunities in this occupation are expected to be good for people who complete automotive training programs in high school, vocational and technical schools, or community colleges. People whose training includes basic electronics skills should have the best opportunities. People without formal mechanic training are likely to face competition for entry-level jobs.

Mechanic careers are attractive to many because they afford the opportunity for good pay and the satisfaction of highly skilled work with one's hands.

Employment opportunities for automotive mechanics are expected to increase about as quickly as the average for all occupations through 2006. Employment growth will continue to be concentrated in automobile dealerships, independent automotive repair shops, and specialty car care chains. Employment of automotive mechanics in gasoline service stations will continue to decline as fewer stations offer repair services.

The number of mechanics will increase because expansion of the driving-age population will increase the number of motor vehicles on the road. The growing complexity of automotive technology necessitates that cars be serviced by skilled workers, contributing to the growth in demand for highly trained mechanics. With more young people entering the job market not interested in mechanics and repair careers, automotive mechanics presents an excellent opportunity for bright, motivated people who have a technical background and desire to make a good living.

More job openings for automotive mechanics are expected than for most other occupations as experienced workers transfer to related occupations, retire, or stop working for other reasons. This large occupation needs a substantial number of entrants each year to replace the many mechanics who leave it.

Most people who enter the occupation can expect steady work because changes in economic conditions have little effect on the automotive repair business. During a downturn, however, some employers may be more reluctant to hire inexperienced workers.

## SALARIES

Median weekly earnings of automotive mechanics who were wage and salary workers were $478 in 1996. The middle 50 percent earned between $333 and $667 a week. The lowest-paid 10 percent earned less than $250 a week, and the top 10 percent earned more than $850 a week.

Many experienced mechanics employed by automotive dealers and independent repair shops receive a commission related to the labor cost charged to the customer. Under this method, weekly earnings depend on the amount of work completed by the mechanic. Employers frequently guarantee commissioned mechanics a minimum weekly salary. Many master mechanics earn from $70,000 to $100,000 annually.

Some mechanics are members of labor unions. The unions include the International Association of Machinists and Aerospace Workers; the International Union, United Automobile, Aerospace and Agricultural Implement Workers of America; the Sheet Metal Workers' International Association; and the International Brotherhood of Teamsters.

## RELATED FIELDS

Other workers who repair and service motor vehicles include diesel truck and bus mechanics; motorcycle mechanics; automotive body repairers, painters, and customizers; and repair service estimators.

## INTERVIEW
### J. G. Barr
## Automotive Mechanic

J. G. Barr is the owner and operator of Poorman's Auto Repair in Cape Coral, Florida. He replaces engines and does tune-ups, brakes, water pumps, and basic general maintenance and repair work. He has been in the field for more than fourteen years.

### How J. G. Barr Got Started

"It all started when I was four, while watching my brother riding a bicycle. It was then I became interested in mechanical things. For Christmas I was given a small toy called Mr. Machine. It was shaped like a mechanical man, with see-through sides that revealed thirty gears inside that made him walk. It could be disassembled and reassembled. This became my favorite toy, and I took it apart many times. I even learned how to change the gear ratio (on my own) to make him walk five times faster than he was designed to.

"At the age of eight, when I was able to repair my go-cart engine inside and out, I decided I wanted to have my own auto repair shop.

"I worked at a steel mill when I turned eighteen, like my dad, his dad, and his dad before. I hated it, but I made such a good living, it was very hard to quit. But fourteen years later, I did just that.

"I left Ohio, moved to Florida, and started my own shop. In just eight months, I was making enough money to build a new house on the water and get the in-ground swimming pool I always wanted. I've given up the pool for flying a private plane now."

## What the Job's Really Like

"As owner of my shop, I typically arrive around 6 A.M. to get an early start before customers arrive, and organize jobs for my employees and myself. I open the shop at 8 A.M. During the day, I answer as many as 100 calls; work on all computer, electrical, and technical repairs; take care of estimates; and attend to customer service. This is a very fast-paced business; my shop is one of the busiest in my city. It's hectic and sometimes frustrating, but rewarding.

"What I like the least about my work is dealing with dissatisfied customers. But the upside is the financial reward and the friends I've made. If I need extra money, I can make it just by working more hours. My employees have the same option. Like most small business owners who have made a success of their profession, I make a decent living."

## Expert Advice

"First, if you don't have a genuine love of cars, don't go into this line of work. There is nothing glamorous about the many scars and burn marks on my hands and arms.

"Auto repair is a rapidly changing profession, so you need to stay educated. Seek out a good school in the automotive field. And find a small, busy shop somewhere, latch onto the owner, and just watch and listen to what he or she says and does. I have trained several employees who have started shops just like mine (one right next door).

"You need to be willing to do whatever it takes. I am forty-five years old, and since I was eight years old I told my family that all I ever wanted to do was own my own shop."

## INTERVIEW
### Roger Pryor
## Transmission Builder

Roger Pryor is the owner and operator of Pryor's Transmission Specialists in Homer, Alaska. He started as a general mechanic and then went into transmission repair. He has an associate's degree in automotive mechanics, has attended various workshops through continuing education, and has a current ASE certification for Master Mechanic.

## How Roger Pryor Got Started

"As a child I loved to take things apart and put them back together to see how they worked. Whenever we would get new parts while I was growing up, I'd take them out of the bag and smell them. I still love the smell of new parts. Personally, I think it's a birth defect.

"I was raised on a farm in southeastern Oklahoma. We didn't have much money, so everything we owned had to be taken care of and repaired. My uncle, who raised me, wasn't very mechanically inclined, so by the time I was eleven or twelve it was my job to make all the repairs. I overhauled my first engine when I was thirteen years old. By the time I was sixteen, people were bringing their problems to me and paying me to fix them.

"It was tough getting my first job at a dealership. I got my two-year degree, and when they asked how much experience I had I told them I was fresh out of school. I hit almost every shop in Houston (where I was at the time), and no one would hire me. Finally I went into a shop—I was twenty years old at the time—and when they asked me how much experience I had, I told them seven years, since that was how long I'd been able to overhaul engines. I haven't been without a job since that day.

"About three years ago, I decided I really didn't want to do general mechanic work any longer. I love transmissions, and that's all I really wanted to work on. I worked in transmission shops as a builder, and last April I opened my own shop. I guess you could say I bought my way into my current job."

## What the Job's Really Like

"As the owner and operator of a small transmission shop, about the only thing I don't do is the office work. My wife does that. My duties include handling technical questions on the phone and diagnosing the vehicle's problem. The first thing I do is determine whether it really is a transmission problem. If it is, then I must determine whether it's internal or external. If it isn't transmission related, I usually suggest a couple of the local shops for the customer to take it to.

"Once I know what the problem is (if it's transmission related), I give the customer an estimate of the cost of labor and parts, as well as the length of time the vehicle will be in the shop.

"I usually don't hunt the parts—that's one of my wife's office duties—but occasionally there is something that she just can't find, or sometimes some of the suppliers are reluctant to deal with a woman. So then I pitch in.

"I try to hire help to remove and reinstall (R&R) the transmission. If help isn't available, I do that also. I do all the building of the transmissions. After the rebuilt transmission is installed in the vehicle, I check underneath to make sure my R&R person got everything set right. Then I take the vehicle for a test drive to make sure the transmission and vehicle are working properly.

"Transmission work requires concentration. When everything works, it's awfully simple; but when it doesn't work, it's simply awful. It is very easy to have a unit rebuilt, then look on your bench and see a snap ring lying there—then you have to take the unit apart and put in the snap ring. Distractions are the most common reason for this. I try to make a mental picture of each step I take so I can visually remind myself that I did put this or that piece in.

"Being orderly and clean are the two most important traits a transmission builder has to have. Dirt is the enemy of a transmission. Your work area has to be clean, and the only parts on your bench should be the parts for the transmission you are currently building.

"A typical day starts at 7:30 A.M. I like to get to work early to make sure everything is clean before the help arrives. I mentally go over what I want each employee to do and what I need to do that day. At about 8:00 A.M., the help starts arriving. I get them going on their jobs and field phone calls and walk-ins until about 8:30. My main duty is to build the transmissions, but since I am the owner, I also handle customers and shop problems.

"Transmission work is usually busy. During the slow time of the year we try to build extra transmissions to set on the shelf, and make any structural improvements needed in the shop. Usually, transmission shops stay busy, though. Up here in Alaska, we don't have a problem with the heat tearing up the transmissions in the summer, but the mountains are hard on them. In the winter, snow and ice can destroy the transmission, so we tend to stay busy most the time.

"Transmission repair is very interesting. In today's market, there are so many different types of transmissions that it really is ever-changing. Some of the most fun is getting my hands on a transmission that I know is going to be abused and building it to handle the abuse.

"When I was working for other shops, I usually worked a forty-hour week, Monday through Friday. Now that I am the owner, my pay has gone down and my hours have been extended.

"What I like most about transmission work is that each transmission has its own set of problems—it never gets boring. I get a sense of personal satisfaction when I test drive a vehicle and the transmission shifts like it's supposed to. I know at that time I've done my job well and the customer will be happy.

"The downside to the business is the outrageous cost of keeping up with the special tools needed and diagnostic equipment that must be upgraded at least every other year.

"In addition, more and more assembly line rebuilders are putting out cheap rebuilds that are inferior in quality but easier to afford. As an independent shop, it's hard to compete with that—we have to justify our prices, but our quality usually does that.

"We are in our first year of business, so right now we are making enough to keep the shop going, pay our employees, and make purchases needed in the ever-changing market. To date, I haven't taken a salary from the shop. It is not uncommon for a transmission shop to gross over \$250,000 a year. The net isn't close to that by the time we pay for parts, labor, rent, utilities, advertising, and all the other expenses related to doing business. An experienced builder should not work for less than \$500 a week. It is not unreasonable for a good, experienced builder to make \$1,000 a week.

"If someone walked into my shop and wanted to learn to build, I'd start him out doing the removing and reinstalling first so he would understand the whole transmission, which will make him a better builder. I usually start unexperienced employees at $7 an hour. As they improve, the pay is raised. Once someone has experience and can put out a quality product in a reasonable amount of time, he can earn $20 to $25 an hour."

## Expert Advice

"Be willing to learn from the bottom up. To build a good transmission, you have to know how the transmission works in the vehicle. If you don't have any experience, accept a job in a transmission shop, even if it's just cleaning the shop. You'll learn a lot just listening to the shop talk going on around you. Learn the different parts. Ask questions.

"The most important qualities you'll need to be successful are cleanliness, orderliness, and accuracy.

"Nothing beats on-the-job training, but most shops will want some formal education also. A degree in automotive mechanics isn't a bad idea, and it will get you a higher starting wage.

"The best way to get started is with your own car. Tinker with it. See how it works. If you can't put your car back together, it might be time to think about a different line of work."

## INTERVIEW
### Harry Forman
## Automotive Mechanic

Harry Forman is the owner of Harry's Auto in West Palm Beach, Florida. He started working on cars at the age of fifteen and has been in the automotive business since 1979. He earned his BBA in 1976 from Temple University in Philadelphia, majoring in accounting.

## How Harry Forman Got Started

"I have always been interested in cars, and at age fifteen I got a job in what was then an Esso (today known as Exxon) gasoline station. I didn't think it would be my profession at that time, as I had planned to be an accountant. I didn't really choose it; it chose me. It is something I am very good at, and it seemed more interesting to me than sitting at a desk all day working as an accountant. I also liked the idea of running my own business.

"I learned how to be a mechanic from on-the-job training. Cars were much simpler at that time, and anyone with a strong mechanical aptitude could probably have become a mechanic.

"I opened my first business in 1979; I created my current job as manager/owner at that time, and it has been with me ever since. There is no job position at my business that I can't fill if one of my employees fails to make it to work. I have passed all the tests required to repair any vehicle in my shop. If I can't fix it, we don't take in the job."

## What the Job Is Really Like

"Most of my day is spent talking to customers, writing estimates, calling for parts, and sweating and toiling while actually fixing cars. I happen to be an excellent diagnostician, which is vital for anyone who wants to be a mechanic/technician.

"When someone brings in a car and describes a shake, rattle, or roll, it is often difficult to figure out exactly what is wrong with the vehicle. A lot of poor mechanics fix everything under the sun and charge an arm and a leg. As a result, mechanics have been tagged with a reputation as rip-off artists. But the problem is often that the mechanics don't know what's wrong with the car and are playing a guessing game, hoping they will eventually figure it out.

"Working in an auto-related field is far from boring. It is fast-paced because everyone wants his or her car back quickly. There is no typical day, but on a busy one the phone begins ringing the moment I put the key in the door and rarely stops.

"Unfortunately, there can be many problems, mishaps, and inconveniences over which I have no control. For example, I call for parts at 10 A.M.; they are promised to me by noon; and after repeated calls, cajoling, and begging, I still don't have parts at

4 P.M. Then the customer calls for the fifth time, and I have to apologize for something that was not my error. Sometimes it takes several days to get all the parts for a job because the auto parts stores do not cooperate.

"There is great satisfaction in being able to fix something that others cannot. Despite all the difficulties, headaches, worries, and time spent trying to get a job to come out right, when the customer is satisfied, I feel very good about myself, the work, and the shop. I have customers who will not go elsewhere, who have rented cars when I was on vacation and waited for me to return. This loyalty is very gratifying. But now that cars have become so complex, people tend to go to a dealer more than a small mom-and-pop shop, and it's more difficult to make as good a living in the business than it once was.

"Salaries range from $7.00 per hour in a depressed area to $27.50 in an affluent section. I own my business, so my income varies weekly depending on conditions. Most shops pay 50 percent of the hourly rate that they bill the customer.

"One final note: the modern mechanic who has up-to-date training prefers to be called a technician."

## Expert Advice

"If you like puzzles, mechanical devices, and this way of thinking, this could be the right work for you. But my advice to someone just starting out who hasn't made the decision already to become a mechanic is to look into becoming a plumber or an electrician. This is mainly because there are no unions for auto mechanics to keep salaries at a rate that keeps up with inflation. Also, it is difficult to run your own shop because of the competition from car dealers and chain store operations."

# FOR MORE INFORMATION

For more details about work opportunities, contact local automotive dealers and repair shops, or the local office of the state employment service. The state employment service also may have information about training programs.

A list of certified automotive mechanic training programs may be obtained from:

National Automotive Technicians Education Foundation
13505 Dulles Technology Drive
Herndon, VA 22071-3415

Information on manufacturer-sponsored two-year associate degree programs in automotive service technology may be obtained from:

Ford ASSET Program
Ford Customer Service Division
Fairlane Business Park III
1555 Fairlane Drive
Allen Park, MI 48101

Chrysler Dealer Apprenticeship Program
National C.A.P. Coordinator
CIMS 423-21-06
26001 Lawrence Avenue
Center Line, MI 48015

General Motors Automotive Service Educational Program
National College Coordinator
General Motors Service Technology Group
MC 480-204-001
30501 Van Dyke Avenue
Warren, MI 48090

Information on how to become a certified automotive mechanic is available from:

ASE
13505 Dulles Technology Drive
Herndon, VA 22071-3415

For general information about the work of automotive mechanics, write:

Automotive Service Association, Inc.
1901 Airport Freeway
Bedford, TX 76021-5732

Automotive Service Industry Association
25 Northwest Point
Elk Grove Village, IL 60007-1035

National Automobile Dealers Association
8400 Westpark Drive
McLean, VA 22102

For a directory of accredited private trade and technical schools that offer programs in automotive technician training, write:

Accrediting Commission of Career Schools and Colleges of
    Technology
2101 Wilson Boulevard, Suite 302
Arlington, VA 22201

For a list of public automotive mechanic training programs, contact:

Vocational Industrial Clubs of America
P.O. Box 3000
1401 James Monroe Highway
Leesburg, VA 22075

# 3

# Motorcycle, Boat, and Small Engine Mechanics

**EDUCATION**
H.S. Preferred

**$$$ SALARY**
$11,600 to $35,000

## OVERVIEW

Although the engines that power motorcycles, boats, and lawn and garden equipment are usually smaller than those that power automobiles and trucks, they have many things in common, including breakdowns. Motorcycle, boat, and small engine mechanics repair and service power equipment ranging from chain saws to yachts.

Small engines, like larger engines, require periodic servicing to minimize the possibility of breakdowns and keep them operating at peak efficiency. At routine intervals, mechanics adjust, clean, and lubricate engines, and, when necessary, replace worn or defective parts such as spark plugs, ignition points, valves, and carburetors. Routine maintenance is normally a major part of the mechanic's work.

When breakdowns occur, mechanics diagnose the cause and repair or replace the faulty parts. The mark of a skilled mechanic is the ability to diagnose mechanical, fuel, and electrical problems and to make repairs in a minimal amount of time. A quick and accurate diagnosis requires problem-solving abilities as well as a thorough knowledge of the equipment's operation. The mechanic first obtains a description of the symptoms of the problem from the owner, and then, if possible, operates the equipment to observe the symptoms. The mechanic may have to use special diagnostic testing equipment and disassemble some components for further examination. After pinpointing the problem, the needed adjustments, repairs, or replacements are made. Some jobs require only the adjustment or replacement of a single item, such as a carburetor or fuel pump, and may be completed in less than an hour. In contrast, a complete engine overhaul may require a number of hours because the mechanic must disassemble and reassemble the engine to replace worn valves, pistons, bearings, and other internal parts.

Motorcycle, boat, and small engine mechanics use common hand tools, such as wrenches, pliers, and screwdrivers, as well as power tools, such as drills and grinders. Engine analyzers, compression gauges, ammeters and voltmeters, and other testing devices help mechanics locate faulty parts and tune engines. Hoists may be used to lift heavy equipment such as motorcycles, snowmobiles, or boats. Mechanics often refer to service manuals for detailed directions and specifications while performing repairs.

Mechanics usually specialize in the service and repair of one type of equipment, although they may work on closely related products. *Motorcycle mechanics* repair and overhaul motorcycles, motorscooters, mopeds, and all-terrain vehicles. Besides engines, they may work on transmissions, brakes, and ignition systems, and make minor body repairs. Because many motorcycle mechanics work for dealers that service only the products they sell, they may specialize in servicing only a few of the many makes and models of motorcycles.

*Motorboat mechanics* repair and adjust the engines and electrical and mechanical equipment of inboard and outboard marine engines. Most small boats have portable outboard engines that can be removed and brought into the repair shop. Larger craft, such as cabin cruisers and commercial fishing boats, are powered by diesel or gasoline inboard or inboard-outboard engines, which are only removed for major overhauls. Motorboat mechanics may also work on propellers, steering mechanisms, marine plumbing, and other boat equipment.

*Small engine mechanics* service and repair outdoor power equipment such as lawnmowers, garden tractors, edge trimmers, and chain saws. They also may occasionally work on portable generators, go-carts, and snowmobiles.

Routine maintenance is a major part of motorcycle mechanics' work.

Motorcycle, boat, and small engine mechanics usually work in repair shops that are well lit and ventilated, but are sometimes noisy when engines are being tested. However, motorboat mechanics may work outdoors in all weather when repairing inboard engines aboard boats; they may have to work in cramped or awkward positions to reach a boat's engine.

In northern states, motorcycles, boats, lawnmowers, and other equipment are used less (or not at all) during the winter, and mechanics may work fewer than forty hours a week; many mechanics are only hired temporarily during the busy spring and summer seasons. Some of the winter slack is taken up by scheduling time-consuming engine overhauls and working on snowmobiles and snowblowers. Many mechanics may work considerably more than forty hours a week when the weather is warmer in the spring, summer, and fall.

# TRAINING

Due to the increasing complexity of motorcycles, most employers prefer to hire motorcycle mechanics who are graduates of formal training programs. However, because technology has not had as great an impact on boat and outdoor power equipment, most boat and small engine mechanics learn their skills on the job. For trainee jobs, employers hire persons with mechanical aptitude who are knowledgeable about the fundamentals of small two- and four-cycle engines. Many trainees develop an interest in mechanics and acquire some basic skills through working on automobiles, motorcycles, boats, or outdoor power equipment as a hobby, or through mechanic vocational training in high school, vocational and technical schools, or community colleges. A growing number also prepare for their careers by completing training programs in motorcycle, marine, or small engine mechanics, but only a relatively small number of such specialized programs exist.

Trainees begin by learning routine service tasks under the guidance of experienced mechanics, such as replacing ignition points and spark plugs or taking apart, assembling, and testing new equipment. Equipment manufacturers' service manuals are an important training tool. As trainees gain experience and proficiency, they progress to more difficult tasks, such as diagnosing the cause of breakdowns or overhauling engines. Up to three years of training on the job may be necessary before an inexperienced beginner becomes skilled in all aspects of the repair of some motorcycle and boat engines.

Employers sometimes send mechanics and trainees to special training courses conducted by motorcycle, boat, and outdoor power equipment manufacturers or distributors. These courses, which can last as long as two weeks, are designed to upgrade the worker's skills and provide information on repairing new models.

Most employers prefer to hire high school graduates for trainee mechanic positions, but will accept applicants with less education if they possess adequate reading, writing, and arithmetic skills. Many equipment dealers employ students part time and during the summer to help assemble new equipment and perform minor repairs. Helpful high school courses include small engine repair, automobile mechanics, science, and business arithmetic.

Knowledge of basic electronics is becoming more desirable for motorcycle, boat, and small engine mechanics. Electronic components are increasingly being used in engine controls, instrument displays, and a variety of other components of motorcycles, boats, and outdoor power equipment. Mechanics should be familiar with at least the basic principles of electronics in order to recognize when an electronic malfunction may be responsible for a problem, and be able to test and replace electronic components.

Motorcycle, boat, and small engine mechanics are sometimes required to furnish their own hand tools. Employers generally provide some tools and test equipment, but beginners are expected to gradually accumulate hand tools as they gain experience. Some experienced mechanics have thousands of dollars invested in tools.

Some mechanics are able to use skills learned through repairing motorcycles, boats, and outdoor power equipment to advance to higher-paying jobs as automobile, truck, or heavy equipment mechanics. In larger shops, mechanics with leadership ability can advance to supervisory positions such as shop supervisor or service manager. Mechanics who are able to raise enough capital may open their own repair shops or equipment dealerships.

## JOB OUTLOOK

Motorcycle, boat, and small engine mechanics held more than 45,000 jobs in North America in 1996. About 12,000 are motorcycle mechanics, while the remainder specialize in the repair of boats or outdoor power equipment such as lawnmowers, garden tractors, and chain saws.

Two-thirds of all motorcycle, boat, and small engine mechanics work for retail dealers of boats, motorcycles, and miscellaneous vehicles. Others are employed by independent repair shops, marinas and boatyards, equipment rental companies, and hardware and lawn and garden stores. About one-fourth are self-employed.

Employment of motorcycle, boat, and small engine mechanics is expected to grow more slowly than the average for all occupations through 2006. The majority of job openings are expected to be replacement jobs because many experienced motorcycle, boat, and small engine mechanics leave each year to transfer to other occupations, retire, or stop working for other reasons. Job prospects should be especially favorable for people who complete mechanic training programs.

Growth of personal disposable income over the 1996–2006 period should provide consumers with more discretionary dollars to buy boats, lawn and garden power equipment, and motorcycles. This will require more mechanics to keep the growing amount of equipment in operation. Also, routine service will always be a significant source of work for mechanics. While technology will lengthen the interval between check-ups, the need for qualified mechanics to perform this service will increase.

Employment of motorcycle mechanics should increase slowly as the popularity of motorcycles rebounds. In the late 1990s, the number of people between the ages of eighteen and twenty-four should begin to grow. Motorcycle usage should continue to be popular with people in this age group, which historically has had the greatest proportion of motorcycle enthusiasts. Motorcycles are also increasingly popular with people over the age of forty. Traditionally, this group has more disposable income to spend on recreational equipment such as motorcycles and boats. Over the next decade, more people will be entering the over-forty age group; this group is responsible for the largest segment of marine craft purchases. These potential buyers will help expand the market for boats while helping to maintain the demand for qualified mechanics.

Construction of new single-family houses will result in an increase in purchase and operation of lawn and garden equipment, increasing the need for mechanics. However, equipment growth will be slowed by trends toward smaller lawns and contracting maintenance to lawn service firms.

## SALARIES

Motorcycle, boat, and small engine mechanics who usually worked full time had median earnings of about $412 a week in 1996. The middle 50 percent earned between $289 and $549 a week. The lowest-paid 10 percent averaged $224 a week, while the highest-paid 10 percent averaged $676 a week.

Motorcycle, boat, and small engine mechanics tend to receive few fringe benefits in small shops, but those employed in larger shops often receive paid vacations, sick leave, and health insurance. Some employers also pay for work-related training and provide uniforms.

## RELATED FIELDS

Motorcycle, boat, and small engine mechanics closely relate to mechanics and repairers who work on other types of mobile equipment powered by internal combustion engines. Related occupations include automotive mechanic, diesel mechanic, farm equipment mechanic, and mobile heavy equipment mechanic.

## INTERVIEW
### Joe Robertson
## Small Engine Mechanic

*Joe Robertson works for Vinton Saw and Mower Service, an independent shop in Vinton, Virginia. He started in 1997 while still in high school and works full time in the summers and part time during the school year.*

## How Joe Robertson Got Started

"I got interested in the small engine a few years after I started mowing grass. I wanted to know how to take better care of my four pieces of outdoor power equipment.

"I took a small engines night course through Roanoke city schools. This was a four-hour, once-a-week class. It lasted twenty weeks or so.

"I got this job through the night course. The instructor offered me the chance to work for him on Saturdays and other days I did not have school."

## What the Job's Really Like

"My duties are to diagnose and repair small air-cooled engines, two- and four-cycles and single and twin cylinders. I also sharpen blades, do winterization service, and do quick fixes for bad spark plugs or clogged carburetor jets. Sometimes I answer the phone and help customers. Mainly, I stay in the back.

"The job can be easy and fun if you make it that way. I always turn on the radio and listen to it all day.

"A typical day starts at 8:00 A.M., pushing out fifty or sixty machines (peak season) and getting the shop ready for ten hours of work. I grab a handful of work orders and go to it. I usually see carburetor problems, internal engine failure, and problems with the outdoor power equipment itself. Sometimes I fix problems as simple as replacing a spark plug. I also see people try to defeat the purpose of the safety mechanisms. This really makes me upset because I have to charge them to replace all the safety stuff, and I would hate for something to happen to someone because they removed the safeguards.

"Around lunchtime, I order pizza or have some other lunch brought to me since I don't really have a lunch hour. I eat for fifteen minutes or so and then continue with the work, sometimes helping customers at the front or on the phone.

"Usually, I try not to stop for anything, so I can get all the work turned out. It can be very busy.

"Ninety-nine times out of one hundred, the job is interesting because you see it all. You never think a lawnmower could get so messed up by a do-it-yourselfer. The only time it could get boring is if you get stuck having to wait on back-ordered parts or you keep coming across the same problem on different machines.

"On a good week, I end up working about fifty-five hours. This is necessary to keep up with the demand and to increase storage space at night. But I can only work full time in the summers. I work part time, about fifteen hours a week, during the school year. I am paid on the amount of work I turn out—for right now anyway. I can make about $150 to $200 on a good week. When I go to college, I will take courses in the computer field. But when I graduate, I still want to work with small engines because I love physical labor. I would not want to sit behind a desk all day working on computers or consulting about computers.

"What I love about this job the most is working with engines. I love the feeling I get when I fix something. I also like going to the technical update seminars in the winter. I get to see all of the new products and learn how the manufacturer is doing and where the company is going. I also like knowing that I can fix almost anything without the help of a shop manual.

"Plus, I have every tool I would ever need to repair anything. (This makes life much easier.) I also like having my boss help me when I have a question.

"What I don't like about the job is some of the customers. Some people will ask if I can get to their machine soon and have it ready. Then when I have it ready for them, they never come to get it. I also don't like it when the customer comes in the back and starts talking to me while I am fixing the equipment. Smoking is also another problem that happens frequently. They come back there and light up, and that smoke mixes with the air and really makes it unbearable. Plus, it is dangerous because of all the gasoline and oil in containers."

## Expert Advice

"If you want to become a small engine mechanic, then get all the training you can from school or vocational education programs. Take a night course if one is offered. If you can work on a car, chances are you can fix a small engine.

"Get started finding a job by looking in the newspapers. You will probably find the most ads for mechanics in the summer. Make sure you mention any training and experience you have. Sometimes the shop will train you if they have time.

"Once on the job, you must be able to work at a moderate pace, repairing about five to six machines a day, depending on the problems. You must also want to get dirty.

"Try to memorize all the sizes of your sockets and wrenches, from one-fourth inch to one inch. This is helpful in doing fast and efficient work. Also, try to remember all the part numbers for common parts such as air filters, carburetor kits, and blades, or know where the parts are located on the shelf. This is really helpful for saving time.

"Buy tools! If your shop will not supply them, then you need to go out and buy lifetime-warranted tools. Get an Impact wrench. This is very helpful when tearing engines down or breaking things loose.

"I have a small engines website at www.lauzonent.com/joeenter. This site mentions a lot of common problems and answers. People also may E-mail me from there with questions."

# FOR MORE INFORMATION

For more details about work opportunities, contact local motorcycle, boat, and lawn and garden equipment dealers, and boatyards and marinas. Local offices of the state employment service may also have information about employment and training opportunities.

General information about motorcycle mechanic careers may be obtained from:

Motorcycle Mechanics Institute
2844 West Deer Valley Road
Phoenix, AZ 85027

American Motorcycle Institute
3042 West International Speedway Boulevard
Daytona Beach, FL 32124

General information about motorboat mechanic careers may be obtained from:

Marine Mechanics Institute
2844 West Deer Valley Road
Phoenix, AZ 85027

American Marine Institute
3042 West International Speedway Boulevard
Daytona Beach, FL 32124

General information about small engine mechanic careers may be obtained from:

Outdoor Power Equipment Institute
341 South Patrick Street
Alexandria, VA 22314

For a list of public motorcycle, boat, and small engine mechanic training programs, contact:

Vocational Industrial Clubs of America
P.O. Box 3000
1401 James Monroe Highway
Leesburg, VA 22075

# CHAPTER 4 Mobile Heavy Equipment Mechanics

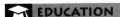

**EDUCATION**
H.S. Preferred

**$$$ SALARY**
$19,000 to $51,000

## OVERVIEW

Mobile heavy equipment is indispensable to construction, logging, surface mining, and other industrial activities. Mobile heavy equipment mechanics service and repair the engines, transmissions, hydraulics, electrical systems, and other components of equipment such as motor graders, trenchers and backhoes, crawler-loaders, and stripping and loading shovels.

Mobile heavy equipment mechanics perform routine maintenance on the diesel engines that power most heavy equipment, and, if an operator reports a malfunction, they search for its cause. First, they inspect and operate the equipment to diagnose the nature of the repairs required. If necessary, they may partially dismantle the engine, examining parts for damage or excessive wear. Then they repair, replace, clean, and lubricate the parts as necessary, and reassemble and test the engine for operating efficiency. If repairs to the drive train are needed, mechanics remove and repair the transmission or differential.

Field service mechanics often work outdoors on construction sites because mobile heavy equipment is too difficult to bring into a repair shop.

Many types of mobile heavy equipment use hydraulics to raise and lower movable parts such as scoops, shovels, log forks, or scraper blades. Repairing malfunctioning hydraulic components is an important responsibility of mobile heavy equipment mechanics. When components lose power, mechanics examine them for hydraulic fluid leaks and replace ruptured hoses or worn gaskets on fluid reservoirs. Occasionally, more extensive repairs are required, such as replacing a defective hydraulic pump.

Mobile heavy equipment mechanics perform a variety of other types of repairs. They diagnose and correct electrical problems and replace defective electronic components. They also disassemble and repair crawler undercarriages and track assemblies. Occasionally, mechanics weld broken body and structural parts, using electric or gas welders.

Many mechanics work in small repair shops of construction contractors, logging and mining companies, and local government road maintenance departments. They typically perform routine maintenance and minor repairs necessary to keep the equipment in operation. Mechanics in larger repair shops— particularly those of mobile heavy equipment dealers and the federal government—perform more difficult repairs, such as rebuilding or replacing engines, repairing hydraulic fluid pumps, or correcting electrical problems. Mechanics in some large shops specialize in one or two types of work, such as hydraulics or electrical systems.

Mobile heavy equipment mechanics use a variety of tools in their work, including common hand tools, such as pliers, wrenches, and screwdrivers; and power tools, such as pneumatic wrenches. They use micrometers and gauges to measure wear on parts, and a variety of testing equipment. For example, they use tachometers and dynamometers to locate engine malfunctions; when working on electrical systems, they use ohmmeters, ammeters, and voltmeters.

Most mobile heavy equipment repair shops are well ventilated, lit, and heated. Many mechanics work indoors in shops, but those who work as field service mechanics spend much of their time away from the shop working outdoors. When mobile heavy equipment breaks down at a construction site, it may be too difficult or expensive to bring it into a repair shop, so a field service mechanic is sent to the jobsite to make repairs. Generally, the more experienced mobile heavy equipment mechanics specialize in field service; they usually drive specially equipped trucks and sometimes must travel many miles to reach disabled machinery. For many mechanics, the independence and challenge of field work outweigh the occasional long hours or bad weather, but other mechanics are more comfortable with the routine of shop work and the opportunity to work as part of a team.

Mechanics handle greasy and dirty parts and often work in awkward or cramped positions. They sometimes must lift heavy tools and parts, and must be careful to avoid burns, bruises, and cuts from hot engine parts and sharp edges of machinery. However, serious accidents may be prevented when the shop is kept clean and orderly and safety practices are observed.

# TRAINING

For trainee jobs, employers hire persons with mechanical aptitude who are high school graduates and at least eighteen years of age. They seek persons knowledgeable about the fundamentals of diesel engines, transmissions, electrical systems, and hydraulics. Although some persons are able to acquire these skills on their own or by working as helpers to experienced mechanics, most employers prefer to hire graduates of formal training programs in diesel or heavy equipment mechanics.

Training programs in diesel and heavy equipment mechanics are given by vocational and technical schools and community and junior colleges. Training in the fundamentals of electronics is also essential because new mobile heavy equipment increasingly features electronic controls and sensing devices. Some one- to two-year programs lead to a certificate of completion; others lead to an associate degree if they are supplemented with additional academic courses. These programs provide a foundation in the basics of diesel and heavy equipment technology, including hydraulics, and enable trainee mechanics to advance more rapidly to the journey, or experienced worker, level.

Through a combination of formal and on-the-job training, trainee mechanics acquire the knowledge and skills to efficiently service and repair the particular types of equipment handled by the shop. Beginners are assigned relatively simple service and repair tasks. As they gain experience and become more familiar with the equipment, they are assigned increasingly difficult jobs and are exposed to a greater variety of equipment.

Many employers send trainee mechanics to training sessions conducted by heavy equipment manufacturers. These sessions, which typically last up to one week, provide intensive instruction in the repair of a manufacturer's equipment. Some sessions focus on particular components found in all of the manufacturer's equipment, such as diesel engines, transmissions, axles, and electrical systems. Other sessions focus on particular types of equipment, such as crawler-loaders and crawler-dozers. As they progress, trainees may periodically attend additional training sessions. Experienced mechanics also occasionally attend training sessions to gain familiarity with new technology or with types of equipment they may never have repaired.

High school courses in automobile mechanics, physics, chemistry, and mathematics provide an essential foundation for a career as a mechanic. Good reading and mathematics skills and a basic understanding of scientific principles are needed to help a mechanic learn important job skills and to keep abreast of new technology through the study of technical manuals. Experience working on diesel engines and heavy equipment acquired in the armed forces is also valuable.

Mobile heavy equipment mechanics usually must buy their own hand tools, although employers furnish power tools and test equipment. Trainee mechanics are expected to accumulate their own tools as they gain experience. Many experienced mechanics have thousands of dollars invested in tools.

Experienced mechanics may advance to field service jobs, where they have greater opportunity to tackle problems independently and earn overtime pay. Mechanics who have leadership ability may become shop supervisors or service managers. Some mechanics open their own repair shops.

# JOB OUTLOOK

Mobile heavy equipment mechanics held about 104,000 jobs in 1996. Nearly 50 percent work for mobile heavy equipment dealers and construction contractors. About 20 percent are employed by federal, state, and local governments; the Department of Defense is the primary federal employer.

Other mobile heavy equipment mechanics worked for surface mine operators, public utility companies, logging camps and contractors, and heavy equipment rental and leasing companies. Still others repaired equipment for machinery manufacturers, airlines, railroads, steel mills, and oil and gas field companies. Fewer than one out of twenty mobile heavy equipment mechanics was self-employed.

Nearly every section of the country employs mobile heavy equipment mechanics in some form, though most work in towns and cities where trucking companies, construction, and other fleet owners have large operations.

Opportunities for heavy equipment mechanics should be good for persons who have completed formal training programs in diesel or heavy equipment mechanics. This is due more to a lack of qualified entrants into the occupation than growth in available jobs. Persons without formal training are expected to encounter growing difficulty entering this occupation.

Employment of mobile heavy equipment mechanics is expected to grow more slowly than the average for all occupations through 2006. Increasing numbers of mechanics will be required to support growth in the construction industry, equipment dealerships, and rental and leasing companies. As equipment becomes more complex, repairs increasingly must be made by specially trained mechanics. More mechanics will be needed by all levels of government to service construction equipment that builds and repairs the country's highways and bridges.

Due to the nature of construction activity, demand for mobile heavy equipment mechanics follows the nation's economic cycle. As the economy expands, construction activity increases, resulting in the use of more mobile heavy equipment. More equipment will be needed to grade construction sites, excavate basements, and lay water and sewer lines, and this will increase the necessity for periodic service and repair. In addition, the construction and repair of highways and bridges also will require more mechanics to service equipment.

Because construction and mining are sensitive to changes in the level of economic activity, mobile heavy equipment may be idled during downturns. In addition, winter is traditionally the slow season for construction activity, particularly in colder regions. Fewer mechanics may be needed during periods when equipment is used less, but employers usually try to retain experienced workers. However, employers may be reluctant to hire inexperienced workers during slow periods.

## SALARIES

Median weekly earnings of mobile heavy equipment mechanics were about $613 in 1996. The middle 50 percent earned from $501 to $762 a week; the lowest 10 percent earned less than $383 a week, and the top 10 percent earned over $981 a week.

Some mobile heavy equipment mechanics are members of unions, including the International Association of Machinists and Aerospace Workers; the International Union of Operating Engineers; and the International Brotherhood of Teamsters.

## RELATED FIELDS

Workers in other occupations who repair and service diesel-powered vehicles and heavy equipment include rail car repairers, and diesel, farm equipment, and mine machinery mechanics.

## INTERVIEW
**Juan Garcia**
Heavy Equipment Mechanic

*Juan Garcia learned his trade as a heavy equipment mechanic during the four years he was enlisted in the U.S. Marine Corps. He* recently finished his tour of duty and will be working for a civilian equipment company in Orlando, Florida.

## How Juan Garcia Got Started

"I enlisted in the marine corps and was picked for this job and trained during my enlistment. After working in the field for a time, I found it enjoyable and fulfilling and decided to continue with it as a civilian career."

## What the Job's Really Like

"Being a heavy equipment mechanic is not a job for someone who does not like getting his or her hands dirty. On a typical day, you might arrive at work to find that a front end loader is down because of a problem with the engine or body, or even hydraulics. You then open your toolbox and get to work.

"First you figure out what is causing the problem, or at least narrow it down to two or three things. Once you diagnose the problem, you then spend most of your time in a tight little place that is hard to reach fixing the problem. That's the part of the work I like the least, working in tight places.

"But when you clean up and realize that you just fixed a very large piece of equipment that lifts, moves, and pushes very heavy things, it's a very good feeling. This job is far from being boring. I really like being able to fix a piece of gear that comes in the shop with a problem that the person operating it doesn't know how to fix or even locate. The work is very satisfying."

## Expert Advice

"Someone who wants to work on heavy equipment should obviously be mechanically inclined. You need to have dedication and commitment, as well as pride in the work you do.

"Most employers are looking for someone who possesses between four and five years of experience. With experience you can expect to earn between $9 to $10 an hour. Someone just starting out shouldn't expect more than about $5.50 to $6.00 an hour."

**INTERVIEW**
**Kirk Woodruff**
Heavy Equipment Mechanic

*Kirk Woodruff worked as a heavy equipment mechanic for garages and lease companies in New Jersey and California for a total of six years. He is now back in college, working toward his bachelor's degree in engineering.*

## How Kirk Woodruff Got Started

"I enjoy fixing things—always have and always will. It feels good to do a clean, professional repair on a truck or trailer and see it back on the road doing what it is designed to do—haul heavy weight over the highways and streets of America.

"My family owned the business I worked for. Watching my father and my uncle is what attracted me to this trade. I got my training on the job and from reading service manuals."

## What the Job's Really Like

"In the beginning, owning a shop meant working eighteen- to twenty-hour days and sleeping in a makeshift bunk house in the shop. I worked there with my dad, uncle, and older brother during my summer vacations in high school. We would leave on Monday and come home late Friday night. Eventually, it turned into a very lucrative business, but not without many years of very hard work.

"After I got out of the army, I went to work full time at the shop. We repaired trucks and trailers that hauled freight all over the nation. We all had to be service managers, cashiers, and so on. Drivers would simply walk into the shop, and we would line up the work on a first come, first served basis. Tire repairs were the heart of the business, so I fixed hundreds of truck tires through the years. I would also pull out transmissions and replace clutches, or let my uncle rebuild the gear box if that was required. I would pull out differentials (rear-end) and rebuild them. I would fabricate metal repairs and weld them in place.

"We all had long-sleeved, heavy coveralls on, and during the summer we would have to take salt pills because we would sweat so much. I was greasy and sweaty twelve to sixteen hours a day in the summer, and was greasy and freezing for twelve to sixteen hours a day during the winter. The only source of heat for many years was a plate steel stove in which we would burn hardwood pallets. It was nice and warm there, but as soon as you walked twenty feet away it was below freezing. I would keep an orchard heater burning near the air compressor in order to keep the entire air pressure system from freezing.

"As a comparison, the shop I worked in recently furnished all of our uniforms with our name on them! They had a nice break room and locker room, too.

"The company had an entire fleet of tractors and trailers that were leased out, so we were busy keeping up with the maintenance on all of the vehicles. I no longer had to do any repairs as we replaced any broken or worn items with new. I did have to rebuild oil coolers and pull radiators and remove and replace clutches.

"There were three eight-hour shifts, running twenty-four hours a day, seven days a week. I was on the graveyard shift, beginning at 10:00 P.M. and ending at 6:30 A.M. Mostly I would do service checks on the vehicles, checking off a long list of items and placing notes for any needed repairs, changing the oil, and checking all fluid levels.

"Now I am attending college with the plan of attaining a degree in engineering. I have many ideas for improvements on heavy mechanical equipment and I hope to market some of them in the future."

## Expert Advice

"My advice to anyone thinking about being a heavy equipment mechanic is to spend some time attending a technical or vocational school or college in that field, because modern trucks are very sophisticated and incorporate a delicate and intricate computer system throughout. Also, there is much to learn about the properties of different types of metal alloys and plastics being produced today that are much more durable than ever before. In fact, it is just about impossible to find a job as a mechanic without some amount of formal training.

"You should enjoy hard work and you should enjoy fixing and repairing things, because that is the job of a mechanic. But be aware that it is fairly hard work and not very clean. They are also saying now that used oil on the skin can cause cancer, as can all of the vapors you inhale on a daily basis."

# FOR MORE INFORMATION

More details about work opportunities for mobile heavy equipment mechanics may be obtained from local mobile heavy equipment dealers, construction contractors, surface mining companies, and government agencies.

Local offices of the state employment service may also have information on work opportunities and training programs.

# CHAPTER 5

# Farm Equipment Mechanics

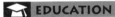
## EDUCATION
H.S. Preferred

## $$$ SALARY
$13,000 to $40,000

## OVERVIEW

Today's farm is typically much larger than in the past, so few if any types of farming can be done economically without specialized machines. Farm equipment has grown in size, complexity, and variety. Many farms have several tractors equipped with 40- to 400-horsepower diesel engines. Self-propelled combines, hay balers, swathers, crop dryers, planters, tillage equipment, grain augers, manure spreaders, and elevators are common, as well as spray and irrigation equipment.

As farm machinery has grown larger with more electronic and hydraulic controls, farmers have increasingly turned to farm equipment dealers for service and repair of the machines they sell. These dealers employ farm equipment mechanics, often called *service technicians*, to do this work and to maintain and repair the smaller lawn and garden tractors many dealers sell to suburban homeowners.

Mechanics spend much of their time repairing and adjusting malfunctioning equipment that has been brought to the shop. But during planting and harvesting seasons, they may travel to farms to make emergency repairs on equipment so that important farming operations are not delayed.

Mechanics also perform preventive maintenance. Periodically, they test, adjust, and clean parts and tune engines. In large shops, mechanics generally specialize in certain types of work, such as diesel engine overhaul, hydraulics, or clutch and transmission repair. Others specialize in repairing the air conditioning units often included in the cabs of combines and large tractors, or in repairing certain types of equipment, such as hay balers. Some mechanics also repair milking, irrigation, and other equipment on farms. In addition, some mechanics who work for dealers and equipment wholesalers assemble new implements and machinery and sometimes do body work, repairing dented or torn sheet metal on tractors or other machinery.

Mechanics use many basic hand tools, including wrenches, pliers, hammers, and screwdrivers. They also use precision equipment, such as micrometers and torque wrenches; engine testing equipment, such as dynamometers, to measure engine performance; and engine analysis units and compression testers, to find worn piston rings or leaking cylinder valves. They use welding equipment or power tools to repair broken parts.

Generally, farm equipment mechanics work indoors. Modern farm equipment repair shops are well ventilated, lit, and heated, but older shops may not offer these advantages. Farm equipment mechanics come in contact with grease, fuel, oil, hydraulic fluid, antifreeze, rust, and dirt, and there is danger of injury when they repair heavy parts supported on jacks or by hoists. Care must also be used to avoid burns from hot engine parts, cuts from sharp edges of machinery, and hazards associated with farm chemicals.

As with most agricultural occupations, the work hours of farm equipment mechanics vary according to the season of the year. During the busy planting and harvesting seasons, mechanics often work six or seven days a week, ten to twelve hours daily. In winter months, however, mechanics may work fewer than forty hours a week, and some may be laid off.

# TRAINING

Farm equipment mechanics must have an aptitude for mechanical work. With the development of more complex farm implements, technical training has become more important. A growing number of employers prefer to hire trainee farm equipment mechanics who have completed a one- or two-year training program in agricultural or diesel mechanics at a vocational or technical school or community or junior college. In general, employers seek persons with training or previous experience in diesel and gasoline engines, the maintenance and repair of hydraulics, and welding, all of which may be learned in many high schools and vocational schools. Mechanics also need a basic knowledge of electronics and must be able to read circuit diagrams and blueprints in order to make complex repairs to electrical and other systems.

Most farm equipment mechanics enter the occupation as trainees and become proficient in their trade by assisting experienced mechanics. The length of training varies with the helper's aptitude and prior experience. At least two years of on-the-job training usually are necessary before a mechanic can efficiently do the more routine types of repair work, and additional training and experience are required for highly specialized repair and overhaul jobs.

Many farm equipment mechanics enter this occupation from a related occupation. For example, they may have experience working as diesel mechanics, mobile heavy equipment mechanics, or automotive mechanics. A farm background is an advantage since working on a farm usually provides experience in basic farm equipment repairs. Persons who enter from related occupations also may start as trainees or helpers, but they may require less on-the-job training.

A few farm equipment mechanics learn the trade by completing an apprenticeship program, which lasts from three to four years and includes on-the-job as well as classroom training in all phases of farm equipment repair and maintenance. Applicants for these programs usually are chosen from shop helpers.

Keeping abreast of changing farm equipment technology requires a great deal of careful study of service manuals and analysis of complex diagrams. Many farm equipment mechanics and trainees receive refresher training in short-term programs conducted by farm equipment manufacturers. These programs usually last several days. A company service representative explains the design and function of equipment and teaches maintenance and repair on new models of farm equipment. In addition, some dealers may send employees to local vocational schools that hold special week-long classes in subjects such as air conditioning repair or hydraulics.

Persons considering a career in this field should have the manual dexterity needed to handle tools and equipment. Occasionally, strength is required to lift, move, or hold heavy parts in place. Difficult repair jobs require problem-solving abilities to diagnose the source of the machine's malfunction. Experienced mechanics should be able to work independently with minimum supervision.

Farm equipment mechanics usually must buy their own hand tools, although employers furnish power tools and test equipment. Trainee mechanics are expected to accumulate their own tools as they gain experience. Experienced mechanics have thousands of dollars invested in tools.

Farm equipment mechanics may advance to shop supervisor, service manager, or manager of a farm equipment dealership. Some mechanics open their own repair shops. A few farm equipment mechanics advance to become service representatives for farm equipment manufacturers.

# JOB OUTLOOK

Farm equipment mechanics held about 44,000 jobs in 1996. Most mechanics work in service departments of farm equipment dealers. Others work in independent repair shops, and in shops on large farms. Most farm equipment mechanics work in small repair shops. Nearly one out of ten farm equipment mechanics is self-employed.

Because nearly every area of the United States has some form of farming, it is common to find farm equipment mechanics employed throughout the country. Employment is concentrated in small cities and towns, making this an attractive career choice for those who wish to live away from a big city. However, many mechanics work in the rural fringes of metropolitan areas, so farm equipment mechanics who prefer the conveniences of city life need not live in rural areas.

Because farms use fewer, but more efficient and reliable, pieces of equipment, employment of farm equipment mechanics should decrease through 2006. Most job openings will arise from the need to replace experienced mechanics who retire. However, opportunities should be good for people who have completed formal training in farm equipment repair or diesel mechanics.

The continued consolidation of farmland into fewer and larger farms and the use of new farming practices means farmers will need a smaller stock of equipment. They will also be more able to invest in new, efficient, and specialized equipment, allowing them to till greater acreage more productively and profitably. For example, new planting equipment uses electronics to spread seeds more uniformly, and electronic controls help harvesters reduce waste.

Farm machinery is expensive and usually designed and manufactured to withstand many years of rugged use. Nevertheless, it requires periodic service and repairs. New farm equipment has longer intervals between service, but because of its increased complexity, many farmers will continue to rely on mechanics for service and repairs. For example, many newer tractors have large, electronically controlled engines and air conditioned cabs, and feature advanced transmissions with many speeds, equipment characteristics that farmers usually cannot repair themselves.

Sales of smaller lawn and garden equipment constitute a growing share of the business of most farm equipment dealers. Most large manufacturers of farm equipment now offer a line of smaller tractors to sell through their established dealerships. However, this equipment is designed for easy home service and requires a mechanic only when major repairs are needed.

The agricultural equipment industry experiences periodic declines—mostly in sales. Layoffs of mechanics, however, are uncommon because farmers often elect to repair old equipment rather than purchase new equipment.

# SALARIES

Farm equipment mechanics had median weekly earnings of about $418 in 1996. The middle 50 percent earned between $312 and $613 a week. The lowest-paid 10 percent earned less than

$256 a week, and the top 10 percent earned over $780 a week. Most farm equipment mechanics also have the opportunity to work overtime during the planting and harvesting seasons, which generally pays time and a half.

Very few farm equipment mechanics belong to labor unions, but those who do are members of the International Association of Machinists and Aerospace Workers; the International Union, United Automobile, Aerospace and Agricultural Implement Workers of America; and the International Brotherhood of Teamsters.

## RELATED FIELDS

Other workers who repair large mobile machinery include aircraft mechanics, automotive mechanics, diesel mechanics, and mobile heavy equipment mechanics.

## INTERVIEW
### Adam Dodds
### Tractor Mechanic

*Adam Dodds has worked for Sterling Farm Equipment, a tractor dealership in Sterling, Ohio, since 1996. He earned his AA degree from Agricultural Technical Institute in Wooster, Ohio, a branch of Ohio State University, majoring in power equipment.*

### How Adam Dodds Got Started

"I enjoy seeing how things like engines and transmissions work. I got started by tinkering on antique farm machinery with my father as a hobby, and I liked it so much that I went to school to learn about more modern machinery and how it works.

"I learned of my job through a friend I go to school with. He is currently employed at Sterling Farm Equipment, and he told me that they needed a mechanic, and that I should talk to the manager. When I went in, he hired me on the spot and I was working by the end of the week."

## What the Job's Really Like

"My duties at Sterling are those that any mechanic would have in a shop. Of course, I am there to fix broken machines, but I also have to keep the shop area clean and help customers with any questions about their own machines.

"I think my job is fun because I get to figure out the problem in a broken machine and fix it with my own two hands. It gives me a lot of satisfaction when the tractor rolls out the doors and is running great.

"I spend my day fixing machines, answering questions, having a good laugh with my co-workers, and reading service bulletins or service manuals. My job is interesting because I always have a variety of things to work on—something new each day.

"I work about fifty hours a week most of the time, and a little more during the busy seasons. I think the work atmosphere is unique at Sterling; it's laid back and relaxed. Now don't get me wrong—there are times where we have to rush to get a tractor done. But in general, the workers, managers, and customers are very easygoing, and that makes it a pleasure to work here.

"I think other shops are a little more strict because they have a flat rate to keep up with and customers to keep happy, but at Sterling we seem to get everything done right and on time, and we don't worry about rushing.

"What I like most is the satisfaction of a job well done. It really feels good to say that machine was broken when it came in, and I fixed it.

"The thing I like least is the pay. I currently earn $7.50 an hour. We get a yearly raise, usually 50 cents per hour and all the overtime we want. I think the amount of work and the type of work deserve a little higher wage; but since I have fun at my job, I don't complain too much."

## Expert Advice

"The best advice I could give someone is to be thorough with your work. Go the extra distance to make sure the job is done right. It will make you and the company look good, showing your commitment to excellent service.

"Also, it is very important to ask questions if you don't know the answer or need help. There is always someone in the shop who will be more than glad to help.

"I think the most important qualities you should possess are open-mindedness and willingness to take suggestions. Most employers have certain ways they like things to be done, so listen to them and learn from what they are trying to teach you.

"As for training, some sort of technical training is helpful, but you really learn most of what you need to know in the shop and from experience. But don't walk into a shop and expect to learn everything you need to know in one day. Experience takes time to gain.

"I think I caught a lucky break getting started, but you just have to keep your eyes open and look for a shop that needs some help. Look in the paper and talk to people in the business. Go into service departments and ask if they need help."

# FOR MORE INFORMATION

Details about work opportunities may be obtained from local farm equipment dealers and local offices of the state employment service.

For general information about the occupation, write to:

Equipment Manufacturers Institute
10 South Riverside Plaza, Room 1220
Chicago, IL 60606

North American Equipment Dealers Association
10877 Watson Road
St. Louis, MO 63127

John Deere and Co.
John Deere Road
Moline, IL 61265

# CHAPTER 6 Elevator Installers and Repairers

**EDUCATION**
H.S. Required

**$$$ SALARY**
$32,000 to $68,000

## OVERVIEW

Elevator installers and repairers—also called *elevator constructors* or *elevator mechanics*—assemble, install, and replace elevators, escalators, dumbwaiters, moving walkways, and similar equipment in new and old buildings. Once the equipment is in service, they maintain and repair it. They are also responsible for modernizing older equipment.

In order to install, repair, and maintain modern elevators, which are almost all electronically controlled, elevator installers and repairers must have a thorough knowledge of electronics, electricity, and hydraulics. Many elevators today are installed with microprocessors, which are programmed to constantly analyze traffic conditions in order to dispatch elevators in the most efficient manner. With these computer controls, it is now possible to get the greatest amount of service with the least number of cars.

When installing a new elevator, elevator installers and repairers begin by studying blueprints in order to determine the equipment layout of the framework to install rails, machines, car enclosures, motors, pumps, cylinders, and plunger foundations. Once the layout analysis is completed, they begin equipment installation. Working on scaffolding or platforms, installers bolt or weld steel rails to the walls of the shaft to guide the elevator up and down.

Elevator installers put in electrical wires and controls by running tubing called *conduit* along the shaft's walls from floor to floor. Mechanics pull plastic-covered electrical wires through the conduit once it's in place. They then install electrical components and related devices required at each floor and at the main control panel in the machine room.

Installers bolt or weld together the steel frame of the elevator car at the bottom of the shaft; install the car's platform, walls, and doors; and attach guide shoes and rollers to minimize the lateral motion of the car as it travels through the shaft. They also install the outer doors and door frames at the elevator entrances on each floor.

For cabled elevators, these workers install geared or gearless machines with a traction drive wheel that guides and moves heavy steel cables connected to the elevator car and counterweight. The counterweight moves in the opposite direction from the car and aids in its swift and smooth movement.

Elevator installers also install elevators in which a car sits on a hydraulic plunger that is driven by a pump. The plunger pushes the elevator car up from underneath, similar to a lift in an auto service station. They also install escalators. They put in place the steel framework, the electrically powered stairs, and the tracks, and install associated motors and electrical wiring. In addition to elevators and escalators, elevator installers also may install devices such as dumbwaiters and material lifts, which are similar to elevators in design, moving walkways, stair lifts, and wheelchair lifts.

The most highly skilled elevator installers and repairers, called *adjusters*, specialize in fine-tuning all of the equipment after installation. Adjusters must make sure that the elevator is working according to specifications, such as stopping correctly at each floor within a specified time period. Once an elevator is operating properly, it must be maintained and serviced regularly to keep it in safe working condition. Elevator maintenance mechanics generally do preventive maintenance, such as oiling and greasing moving parts, replacing worn parts, testing equipment with meters and gauges, and adjusting equipment for optimal performance. They also troubleshoot and may be called in to do emergency repairs.

A service crew usually handles major repairs—for example, replacing cables, elevator doors, or machine bearings. This may require cutting torches or rigging equipment—tools a maintenance mechanic would not normally carry. Service crews also do major modernization and alteration work such as moving and replacing electrical motors, hydraulic pumps, and control panels.

Elevator installers and repairers usually specialize in installation, maintenance, or repair work. Maintenance and repair workers generally need more knowledge of electricity and electronics than installers because a large part of maintenance and repair work is troubleshooting. Similarly, construction adjusters need a thorough knowledge of electricity, electronics, and computers to ensure that newly installed elevators operate properly.

Most elevator installers and repairers work a forty-hour week. However, maintenance and service mechanics often work overtime when repairing essential elevator equipment. They are sometimes on twenty-four–hour call. Maintenance mechanics, unlike most elevator installers, are on their own most of the day and typically service the same elevators periodically.

Elevator installers lift and carry heavy equipment and parts and may work in cramped spaces or awkward positions. Hazards include falls, electrical shock, muscle strains, and other injuries related to handling heavy equipment. Because most of their work is performed indoors in buildings under construction or in existing buildings, elevator installers and repairers lose less work time due to inclement weather than other building trades workers.

## TRAINING

Most elevator installers and repairers apply for their jobs through a local of the International Union of Elevator Constructors. Applicants for trainee positions must be at least eighteen years old, have a high school diploma or equivalent, and pass an aptitude test. Good physical condition and mechanical skill also are important.

Elevator installers and repairers learn their trade in a program administered by local joint educational committees representing the employers and the union. These programs, through which the trainee learns everything from installation to repair, combine on-the-job training with classroom instruction in electrical and electronic theory, mathematics, applications of physics, and safety. Elevator installers and repairers in nonunion shops may complete training programs sponsored by independent contractors.

Generally, trainees or helpers must complete a six-month probationary period. After successful completion, they work toward becoming fully qualified mechanics within four to five years. In order to be classified a fully qualified mechanic, union trainees must pass a standard mechanics examination administered by the National Elevator Industry Educational Program. Most states and cities also require elevator constructors to pass a licensing examination.

Most trainees or helpers assist experienced elevator installers and repairers. Beginners carry materials and tools, bolt rails to walls, and assemble elevator cars. Eventually, they learn to do more difficult tasks, such as wiring, which requires a knowledge of local and national electrical codes.

High school courses in electricity, mathematics, and physics provide a useful background. As elevators become increasingly sophisticated, workers may find it necessary to acquire more advanced formal education—for example, in postsecondary technical school or junior college—with an emphasis on electronics. Workers with more formal education generally advance more quickly than their counterparts.

Many elevator installers and repairers also receive training from their employers or through manufacturers to become familiar with the company's particular equipment. Retraining is very important to keep abreast of technological developments in elevator repair. In fact, union elevator constructors typically receive continual training throughout their careers through correspondence courses, seminars, or formal classes. Although voluntary, this training greatly improves one's chances for promotion.

Some installers may receive further training in specialized areas and advance to mechanic-in-charge, adjuster, supervisor, or elevator inspector. Adjusters, for example, may actually be picked for the position because they possess particular skills or are seen to be more electronically inclined. Others workers may move into management, sales, or product design.

## JOB OUTLOOK

Elevator installers and repairers held about 25,000 jobs in 1996. Most are employed by special trade contractors. Others are employed by field offices of elevator manufacturers; wholesale distributors; small local elevator maintenance and repair contractors; or by government agencies or businesses that do their own elevator maintenance and repair.

Employment of elevator installers and repairers is expected to increase more slowly than the average for all occupations through 2006, and relatively few new job opportunities will be generated because the occupation is small. Replacement needs, another source of jobs, also will be relatively low, in part because a substantial amount of time is invested in specialized training that yields high earnings so that workers tend to remain in the field.

The job outlook for new workers is largely dependent on activity in the construction industry, and opportunities may vary from year to year as conditions within the industry change. Job prospects should be best for those with postsecondary training in electronics or more advanced formal education.

Demand for elevator installers and repairers will increase as equipment ages and needs more repairs and as the construction of new buildings with elevators and escalators increases. Growth also should be driven by the need to continually update and modernize older equipment, including improvements in appearance and the installation of more sophisticated equipment and computerized controls. Because equipment must always be kept in working condition, economic downturns will have less of an effect on employment of elevator maintenance and repair mechanics than on other occupations. The need for people to service elevators and escalators should increase as equipment becomes more intricate and complex.

## SALARIES

Median weekly earnings of elevator installers and repairers who worked full time were $844 in 1996. The middle 50 percent earned between $740 and $1,088. The lowest 10 percent earned less than $633 a week, and the top 10 percent earned more than $1,322 a week.

Average weekly earnings for union elevator installers and repairers were about $865 in 1996, according to data from the International Union of Elevator Constructors. Rates vary with geographic location. Probationary helpers started at about 50 percent of the rate for experienced elevator mechanics, or about $432 a week. Nonprobationary helpers earned about 70 percent of this rate, or an average of about $605 a week. Mechanics-in-charge averaged $973 a week.

In addition to free continuing education, elevator installers and repairers receive basic benefits enjoyed by most other workers.

The proportion of elevator installers and repairers who are union members is higher than in nearly any other occupation. Almost 80 percent of elevator installers and repairers are members of the International Union of Elevator Constructors, compared to 15 percent in all occupations and 23 percent for other craft and repair occupations.

## RELATED FIELDS

Elevator installers and repairers combine electrical and mechanical skills with construction skills such as welding, rigging, measuring, and blueprint reading. Other occupations that require many of these skills are boilermaker, electrician, industrial machinery repairer, millwright, sheet-metal worker, and structural ironworker.

# INTERVIEW
## Pete Bennett
## Elevator Mechanic

*Pete Bennett is an elevator mechanic with Westcon Elevator, an elevator service, repair, and installation company in Fountain Valley, California. He's been working in this field since 1980.*

## How Pete Bennett Got Started

"It was fate that I got involved with this work. I met a man at a Christmas party, and he told me that the elevator union hall needed help. I was building houses and starting a family in Denver, and the house-building economy went bad. Three months later, I was installing elevators in downtown Denver. It was a boom time, and my building skills lent themselves to installing elevators.

"But then elevator construction dried up in Denver. I was offered a ground-level nonunion job in southern California with a brand-new company. That was thirteen years ago, and I've worked for the same company since.

"Most of my training was hands-on and practical. The union did require schooling to become a journeyman."

## What the Job's Really Like

"The job I have is wide and varied. I am responsible for safe elevators and happy customers. A big part of my work is doing safety checks and preventive maintenance. I check lightbulbs; I check safe circuits; I check motor starter contacts to prevent catastrophic motor failure. All of this is to ensure proper elevator operation.

"I work mostly with hydraulic elevators, and some cable elevators. I have installed more than 400 elevators in the last twelve years, and I check on probably 45 to 50 elevators a month. My company's goal is to have a hassle-free operation. We want to go in, do our work quietly, finish up, and get out of there. People get worried if they see an elevator under repair.

"A large part of the job is public relations, reassuring customers that the elevator is fine and that they have nothing to worry about.

"My job involves a lot of driving. In an average day, I get paid for eight hours and work thirteen, including the drive, which is at least 100 miles a day. I also have to fight the traffic in southern California.

"Sometimes I have to stay overnight when sent on jobs out of town. With most new construction, the town doesn't necessarily have any experienced elevator people, so they have to borrow them from other cities. I liked the traveling at first, but that positive feeling can fade quickly. Every hotel room looks the same after a while.

"The routine is what I like least. If all I did was service elevators, I'd be bored. Doing preventive maintenance is like brushing your teeth, so I don't get excited about it.

"I prefer the public relations aspect of my job. I reassure people frequently, and I get to meet different people all the time. I like being able to solve problems and use my people skills. I also like to remodel the interior of elevators. I get to do that eight or ten times a year. I have carpenter skills. I work with formica, mirrors, and wood.

"If you put in a good product, you might make some money—or not, depending on the times. The goal is to get the elevator on a full-service maintenance contract: The customer pays a monthly charge, and I visit each elevator monthly and spend an hour a month with each one. If I don't take care of it and something breaks, then I have to pay to fix it."

## Expert Advice

"Strong electrical troubleshooting skills are needed to advance in the elevator trade. The armed forces is a good place to get the skills. Vocational and technical schools are another good way to go. It's a good job for someone who's young and mechanically inclined.

"The salary varies greatly between union and nonunion. Here in L.A., journeyman wages are $31.02 per hour. The best bet for someone just starting out is to have electrical troubleshooting knowledge. A probationary helper makes 50 percent of a journeyman's wage. After six months, he is eligible for the helpers' test. If he passes, his wage goes to 70 percent of a journeyman. Elevator service and repair work is steadier work. To make the transition from probationary construction helper to journeyman service technician requires strong electrical troubleshooting skills.

"Good people skills are also required to keep building managers happy. And you have to be self-motivated to be a service worker. Working by yourself, it's easy to be tempted to goof off. I could easily do that, but I have too much work to do to even think about it."

# FOR MORE INFORMATION

For further details about opportunities as an elevator installer and repairer, contact elevator manufacturers, elevator repair and maintenance contractors, a local of the International Union of Elevator Constructors, or the nearest local public employment service office.

# CHAPTER 7

# Home Appliance and Power Tool Repairers

 **EDUCATION**
H.S. Required

**$$$ SALARY**
$13,000 to $40,000

## OVERVIEW

Appliance and power tool repairers, often called *service technicians*, repair home appliances such as ovens, washers, dryers, refrigerators, window air conditioners, and vacuum cleaners, as well as power tools such as saws and drills. Some repairers only service small appliances such as microwaves and vacuum cleaners; others specialize in major appliances such as refrigerators, dishwashers, washers, and dryers; and others only handle power tools or gas appliances.

To determine why an appliance or power tool fails to operate properly, repairers visually inspect it and run it to check for unusual noises, excessive vibration, fluid leaks, or loose parts. They may have to consult service manuals and troubleshooting guides to diagnose particularly difficult problems. They may disassemble the appliance or tool to examine its internal parts for signs of wear or corrosion. To check electrical systems for shorts and faulty connections, repairers follow wiring diagrams and use testing devices such as ammeters, voltmeters, and wattmeters.

After identifying problems, they replace or repair defective belts, motors, heating elements, switches, gears, or other items. They tighten, align, clean, and lubricate parts as necessary. Repairers use common hand tools including screwdrivers, wrenches, files, and pliers, as well as soldering guns and special tools designed for particular appliances. When servicing appliances with electronic parts, they may replace circuit boards or other electronic components.

When servicing refrigerators and window air conditioners, repairers must use care to conserve, recover, and recycle chlorofluorocarbon (CFC) and hydrochlorofluorocarbon (HCFC) refrigerants used in their cooling systems. The release of CFCs and HCFCs is thought to contribute to the depletion of the stratospheric ozone layer, which protects plant and animal life from ultraviolet radiation. Repairers conserve the refrigerant by making sure that there are no leaks in the system; they recover it by venting the refrigerant into proper cylinders; and they recycle it for reuse with special filter-dryers.

Repairers servicing gas appliances may check the heating unit and replace pipes, thermocouples, thermostats, valves, and indicator spindles. They also answer emergency calls for gas leaks. To install gas appliances, they may have to install pipes in customers' homes to connect the appliances to the gas line. They measure, lay out, cut, and thread pipe and connect it to a feeder line and to the appliance.

They may have to saw holes in walls or floors and hang steel supports from beams or joists to hold gas pipes in place. Once the gas line is in place, they turn on the gas and check for leaks.

Repairers also answer customers' questions about the care and use of appliances. For example, they demonstrate how to load automatic washing machines, arrange dishes in dishwashers, or sharpen chain saws.

Repairers write up estimates of the cost of repairs for customers, keep records of parts used and hours worked, prepare bills, and collect payment.

Home appliance and power tool repairers who handle portable appliances usually work in repair shops that generally are quiet, well lit, and adequately ventilated. Those who repair major appliances usually make service calls to customers' homes. They carry their tools and a number of commonly used parts with them in a truck or van and may spend several hours a day driving. They may work in clean, comfortable rooms such as kitchens, but sometimes the appliance is in an area of the home that is damp, dirty, or dusty. Repairers sometimes work in cramped and uncomfortable positions when replacing parts in hard-to-reach areas of appliances.

Repairer jobs generally are not hazardous, but service technicians must exercise care and follow safety precautions to avoid electrical shocks and injuries when lifting and moving large appliances. When servicing gas appliances and microwave ovens, they must be aware of the dangers of gas and radiation leaks.

Many home appliance and power tool repairers work a standard forty-hour week. Some work early mornings, evenings, and Saturdays. During hot weather, repairers of air conditioners and refrigerators are in high demand by consumers, and many work overtime. Repairers of power tools such as saws and drills may also have to work overtime during spring and summer months, when use of such tools increases and breakdowns are more frequent.

Home appliance and power tool repairers usually work with little or no direct supervision, a feature of the job that appeals to many people.

# TRAINING

Employers generally require a high school diploma for home appliance and power tool repairer jobs. Many repairers learn the trade primarily on the job. Mechanical aptitude is desirable, and those who work in customers' homes must be courteous and tactful.

Employers prefer to hire people with formal training in appliance repair and electronics, and many repairers complete one- or two-year formal training programs in appliance repair and related subjects in high schools, private vocational schools, and community colleges. Courses in basic electricity and electronics are becoming increasingly necessary as more manufacturers are installing circuit boards and other electronic control systems in home appliances.

Regardless of whether their basic skills are developed through formal training or on the job, trainees usually get additional training from their employer. In shops that fix portable appliances, trainees work on a single type of appliance, such as vacuum cleaners, until they master its repair. Then they move on to others, until they can repair all those handled by the shop. In companies that repair major appliances, beginners assist experienced repairers on service visits. They may also study on their own. They learn to read schematic drawings, analyze problems, determine whether to repair or replace parts, and follow proper safety procedures. Up to three years of on-the-job training may be needed to become skilled in all aspects of repair of the more complex appliances.

Some appliance and power tool manufacturers and department store chains have formal training programs that include home study and shop classes in which trainees work with demonstration appliances and other training equipment. Many repairers receive supplemental instruction through two- or three-week seminars conducted by appliance and power tool manufacturers. Experienced repairers also study service manuals and attend training classes.

The Environmental Protection Agency (EPA) has mandated that all repairers who purchase or work with refrigerants must be certified in its proper handling. To become certified to purchase and handle refrigerants, repairers must pass a written examination. Exams are administered by organizations approved by the Environmental Protection Agency, such as trade schools, unions, and employer associations. Though no formal training is required for certification, many of these organizations offer training programs designed to prepare workers for the certification examination.

To protect consumers, some states and areas require repairers to be licensed or registered. Applicants for licensure must meet standards of education, training, and experience; they also may have to pass an examination, which can include a written examination, a hands-on practical test, or a combination of both.

Repairers in large shops or service centers may be promoted to supervisor, assistant service manager, or service manager. A few advance to managerial positions such as regional service manager or parts manager for appliance or tool manufacturers. Preference is given to those who demonstrate technical competence and show an ability to get along with co-workers and customers. Experienced repairers who have sufficient funds and knowledge of small business management may open their own repair shops.

## JOB OUTLOOK

Home appliance and power tool repairers held nearly 71,000 jobs in 1996. Nearly one out of ten repairers is self-employed. Almost two out of three salaried repairers work in retail establishments such as department stores, household appliance stores, and fuel dealers. Others work for gas and electric utility companies, electrical repair shops, and wholesalers.

Almost every community in the country employs appliance and power tool repairers; a high concentration of jobs is found in more populated areas.

Employment of home appliance and power tool repairers is expected to increase more slowly than the average for all occupations through 2006. The number of home appliances and power tools in use is expected to increase as the number of households and businesses grows and new and improved appliances and tools are introduced. The increased use of electronic parts such as solid-state circuitry, microprocessors, and sensing devices in appliances will reduce the frequency of repairs. Nevertheless, as the current pool of appliance and power tool repairers retire or transfer to other occupations, job openings will arise. Prospects should continue to be good for well-trained repairers, particularly those with a strong background in electronics. Most people with the electronics training needed to repair appliances go into other repairer occupations.

Employment is relatively steady because the demand for appliance repair services continues even during economic downturns. Jobs are expected to be increasingly concentrated in larger companies as the number of family-owned businesses and smaller shops decline.

## SALARIES

Home appliance and power tool repairers who usually worked full time had median earnings of $579 a week in 1996. The middle 50 percent earned between $354 and $760 a week. The lowest-paid 10 percent earned $255 a week or less, while the highest-paid 10 percent earned $929 a week or more. Earnings of home appliance and power tool repairers vary widely according to the skill level required to fix equipment, geographic location, and the type of equipment repaired. Earnings tend to be highest in large firms and for those servicing gas appliances. Many receive commission along with their hourly wage.

Many larger dealers, manufacturers, and service stores offer benefits such as health insurance coverage, sick leave, and retirement and pension programs. Some home appliance and power tool repairers belong to the International Brotherhood of Electrical Workers.

# RELATED FIELDS

Other workers who repair electrical and electronic equipment include heating, air conditioning, and refrigeration mechanics; pinsetter mechanics; office machine and cash register servicers; electronic home entertainment equipment repairers; and vending machine servicers and repairers.

# INTERVIEW
## Michael Sanchez
## Owner of a Power Tool Sales and Repair Business

Michael Sanchez is part owner of Rio Industrial Supply Company, a construction sales outfit in El Paso, Texas. He's been in the business since 1979.

## How Michael Sanchez Got Started

"I grew up maintaining some apartments with my grandfather, so I was always taking something apart. I was almost out of high school and I had been working for the family since I was about ten years old. When one of my dad's employees quit, he needed someone to fill the position, and I already had some experience taking tools apart. (I just always had a heck of a time putting them back together.)

"The most training I really ever had was from a Bosch Factory Service Center seminar. It was two days long and very intense in both paperwork and hands-on training. The hands-on part was my favorite. This has helped me in both the repair and sale of the tools, because I know which tools are good and which aren't."

## What the Job's Really Like

"I start my week by ordering what we needed last week in parts. This includes special orders, parts we have run out of, and parts that I feel we might need in the future.

"Let me explain about ordering parts we might need in the future. Makita, for instance, comes out with an average of twenty-five new tools each year. You need to figure out which parts you will need right away. It's usually a safe bet to order switches, brushes, cords (if they're a special kind), and at least one housing set. Most power tools will get dropped in their first year, because the guy who buys it will take that tool to its extreme limits to see how well it holds up before he buys another one, and you better have a housing if he drops it. This is not covered under warranty, thus the customer incurs a small charge. Hmm, I hear my register ringing up a sale—my heart is in tool repairs—but daddy didn't raise a fool; I have to find a way to make up for my salary.

"Next, I look up the technical bulletins and put into my manuals the latest updates for new and old tools. There is an average of 100 updates a month from all the tool manufacturers we represent.

"Then it's out to the streets, hitting the pavement looking for sales and calling on my regulars. I spend 90 percent of my outside sales time across the border in Mexico. I call on plants that manufacture anything from the seat you sit on in your car or truck to artificial Christmas trees.

"Back at the office that afternoon, I line up the tools that have come in for repair, both warranty and nonwarranty.

"I also order the accessories for most of the tools we sell. I do this four different ways. First, I look at the new catalogs that come out. Second, I read a lot of magazines that show new goodies coming out (a new drill bit or battery, for example) and the reviews they get from different groups such as the American Woodsmen. Third, we get a lot of weekend warriors in our store; these guys love their tools and will give you all kinds of advice about new and upcoming stuff, or what they wish was available to them. Last, we get a lot of phone-in orders, and I have to keep up with what these people want in both parts and accessories.

"We have a philosophy—don't order just one of something if you can order several extras. This is how I have built up my huge inventory of parts. If a customer doesn't buy it today, another one might need it eventually.

"Tuesday and Thursday are my days in, the days when no one bothers me in my workroom. I fix as many tools as possible that day and make up their invoices or warranty forms as needed. A lot of people call asking about how to fix this or that on a tool. Like a doctor, I make no money by giving out free advice, but if it's a really good customer or friend I will help to a certain point. I try to keep people from fixing their own tools, which keeps people from getting hurt—electrocuted or losing a finger, for example—and helps the customer get the most use out of his tools.

"What I consider an upside to my work someone might say is corny: but it's the look I see on a customer's face when I have fixed his tools and he doesn't have to buy more for a while.

"Of course, collecting money earned for the hard work I put into repairing these tools is good. My customers know that I'm not cheap, but I fix their tools right the first time and they don't have to worry about one of their employees getting hurt using a tool I fixed.

"The downsides—first, you have the guys who bring their sick tool in a basket and you play twenty questions, trying to figure out what it is—a drill, a screwdriver, or what.

"Second, I guess I feel for tools. I hate more than anything to see an abused tool. I've seen everything from a drill with a broken housing that was being held together with wire and duct tape, to a saw on which the owner riveted all the loose and broken parts. He wanted to know if I could drill out the rivets and put the correct screws back on. (I hear my register ringing!)

"Everyone wants you to warranty their tools. The manufacturers we represent won't admit it, but they love us. We keep their costs down compared to a large chain store. These stores help the guy who wants to use a tool once, take it back, and get his money refunded. Then the tool manufacturers look bad because the stores see certain tools being brought back frequently, even though 90 percent of them have nothing wrong."

## Expert Advice

"Be prepared to listen a lot; you learn more by listening than talking. And you should read, read, read—and I mean a lot. There's a whole new tool out there being built as we speak, and you may not know how to fix it if you don't know anything about it.

"Treat everyone you talk to at the tool manufacturer with the utmost respect; they will help you with the best sales tool, word of mouth. These people will send business your way because they know you are doing things correctly and know what you must do to keep their brand of tools going. This makes them look good because the customer will stick to a brand if he knows you give it a good review and can always repair it for him.

"As for training, if you get invited to go to a class or seminar for tool repair, make sure you go. Manufacturers always are giving classes to fix their tools; they want you to be informed about their brand in particular (we fix almost every brand out there).

"Most important—try to really put in an effort at the hands-on seminars and classes. Tool companies love to see someone who's not afraid to get elbow deep into a tool. I should know—I've been called on by several companies to work for them or in the development of their new tools."

# FOR MORE INFORMATION

For information about jobs in the home appliance and power tool repair field, contact local appliance repair shops, manufacturers, vocational trade schools, appliance dealers, and utility companies, or the local office of the state employment service.

For general information about the work of home appliance repairers, contact:

Appliance Service News
P.O. Box 789
Lombard, IL 60148

National Association of Service Dealers
P.O. Box 9680
Denver, CO 80222

United Servicers Association, Inc.
P.O. Box 59707
Dallas, TX 75229

National Appliance Service Association
9247 N. Meridian, Suite 216
Indianapolis, IN 46260

For information about technician certification, as well as general information about the work of home appliance repairers, contact:

National Appliance Service Technician Certification
    Program (NASTeC)
20 N. Wacker Drive, Suite 1231
Chicago, IL 60606

Professional Service Association
71 Columbia Street
Cohoes, NY 12047

National Association of Service Dealers
10 E. 22nd Street, Suite 310
Lombard, IL 60148

# CHAPTER 8

# Electronic Equipment Repairers

**EDUCATION**
H.S. Required

**$$$ SALARY**
$17,000 to $50,000

## OVERVIEW

Electronic equipment repairers, also called *service technicians* or *field service representatives*, install, maintain, and repair electronic equipment used in offices, factories, homes, hospitals, aircraft, and other places. Equipment includes televisions, radar, industrial equipment controls, computers, telephone systems, and medical diagnosing equipment. Repairers have numerous job titles, which often refer to the kind of equipment they work with.

Electronic repairers install, test, repair, and calibrate equipment to ensure that it functions properly. They keep detailed records on each piece of equipment to provide a history of tests, performance problems, and repairs.

When equipment breaks down, repairers first examine work orders, which indicate problems, or talk to equipment operators. Then they check for common causes of trouble such as loose connections or obviously defective components. If routine checks do not locate the trouble, repairers may refer to schematics and manufacturers' specifications that show connections and provide instruction on how to locate problems. They use voltmeters, ohmmeters, signal generators, ammeters, and oscilloscopes and run diagnostic programs to pinpoint malfunctions. It may take several hours to locate a problem, but only a few minutes to fix it. However, more equipment now has self-diagnosing features that greatly simplify the work. To fix equipment, repairers may replace defective components, circuit boards, or wiring, or adjust and calibrate equipment, using test equipment, small hand tools such as pliers and screwdrivers, and soldering irons.

Field repairers visit worksites in their assigned area on a regular basis to do preventive maintenance according to manufacturers' recommended schedules and whenever emergencies arise. During these calls, repairers may also advise customers on how to use equipment more efficiently and how to spot problems in their early stages. They also listen to customers' complaints and answer questions, promoting customer satisfaction and goodwill. Some field repairers work full time at installations of clients with a lot of equipment.

Bench repairers work at repair facilities, stores, factories, or service centers. They repair portable equipment, such as televisions and personal computers brought in by customers, or defective components and machines requiring extensive repairs that have been sent in by field repairers. They determine the source of a problem in the equipment, and may estimate whether it is wiser to buy a new part or machine or to fix the broken one.

Some electronic equipment repairers work shifts, including weekends and holidays, to service equipment in computer centers, manufacturing plants, hospitals, and telephone companies that operate around the clock. Shifts are generally assigned on the basis of seniority. Repairers may also be on call at any time to handle equipment failure.

Repairers generally work in clean, well-lit, air conditioned surroundings—an electronic repair shop or service center, hospital, military installation, or a telephone company's central office. However, some workers, such as commercial and industrial electronic equipment repairers, may be exposed to heat, grease, and noise on factory floors. Some may have to work in cramped spaces. Telephone installers and repairers may work on rooftops, ladders, and telephone poles.

The work of most repairers involves lifting, reaching, stooping, crouching, and crawling. Adherence to safety precautions is essential to guard against work hazards such as minor burns and electrical shock.

# TRAINING

Most employers prefer applicants with formal training in electronics. Electronic training is offered by public postsecondary vocational and technical schools, private vocational schools and technical institutes, junior and community colleges, and some high schools and correspondence schools. Programs take one to two years. The military services also offer formal training and work experience.

Training includes general courses in mathematics, physics, electricity, electronics, schematic reading, and troubleshooting. Students also choose courses that prepare them for a specialty, such as computers, commercial and industrial equipment, or home entertainment equipment. A few repairers complete formal apprenticeship programs sponsored jointly by employers and locals of the International Brotherhood of Electrical Workers.

Applicants for entry-level jobs may have to pass tests that measure mechanical aptitude, knowledge of electricity or electronics, manual dexterity, and general intelligence. Newly hired repairers, even those with formal training, usually receive some training from the employer. They may study electronics and circuit theory and math. They also get hands-on experience with equipment, doing basic maintenance and using diagnostic programs to locate malfunctions. Training may be in a classroom or it may be self-instruction, consisting of videotapes, programmed computer software, or workbooks that allow trainees to learn at their own pace.

Experienced technicians attend training sessions and read manuals to keep up with design changes and revised service procedures. Many technicians also take advanced training in a particular system or type of repair.

Good eyesight and color vision are needed to inspect and work on small, delicate parts, and good hearing to detect malfunctions revealed by sound. Because field repairers usually handle jobs alone, they must be able to work without close supervision. For those who have frequent contact with customers, a pleasant personality, neat appearance, and good communications skills are important. Repairers must also be trustworthy, because they may be exposed to money and other valuables in places like banks and securities offices; some employers require that they be bonded. A security clearance may be required for technicians who repair equipment or service machines in areas where people are engaged in activities related to national security.

The International Society of Certified Electronics Technicians and the Electronics Technicians Association each administer a voluntary certification program. In both, an electronics repairer with four years of experience may become a certified electronics technician. Certification, which is by examination, is offered in computer, radio-TV, industrial and commercial equipment, audio, avionics, wireless communications, video distribution, satellite, and radar systems repair. An associate-level test, covering basic electronics, is offered for students or repairers with less than four years of experience. Those who test and repair radio transmitting equipment, other than business and land mobile radios, need a general operators license from the Federal Communications Commission.

Experienced repairers with advanced training may become specialists or troubleshooters who help other repairers diagnose difficult problems, or work with engineers in designing equipment and developing maintenance procedures.

Because of their familiarity with equipment, repairers are particularly well qualified to become manufacturers' sales workers. Workers with leadership ability also may become maintenance supervisors or service managers. Some experienced workers open their own repair services or shops, or become wholesalers or retailers of electronic equipment.

# JOB OUTLOOK

Electronic equipment repairers held about 396,000 jobs in 1996. Many work for telephone companies. Others work for electronic and transportation equipment manufacturers, machinery and equipment wholesalers, hospitals, electronic repair shops, and firms that provide maintenance under contract (called third-party maintenance firms). The distribution of employment by occupation was as follows:

| | |
|---|---|
| Computer and office machine repairers | 141,000 |
| Communications equipment mechanics | 116,000 |
| Commercial and industrial electronic equipment repairers | 60,000 |
| Telephone installers and repairers | 37,000 |
| Electronic home entertainment equipment repairers | 33,000 |

Overall, employment of electronic equipment repairers is expected to grow more slowly than the average for all occupations through 2006. Although the amount of electronic equipment in use will grow very rapidly, improvements in product reliability and ease of service and lower equipment prices will dampen the need for repairers.

Employment of computer equipment repairers will grow much more rapidly than average for all occupations through 2006 as the number of computers in service increases rapidly. Employment of commercial and industrial equipment repairers outside the federal government will increase more quickly than the average as the amount of equipment grows. Mainly because of cuts in the defense budget, their employment in the federal government will decline. Employment of those who repair electronic home entertainment equipment will decline as equipment becomes more reliable and easier to service.

Telephone installer jobs are expected to decline sharply, and jobs for communication equipment mechanics are expected to grow more slowly than the average because of improvements in telephone equipment reliability, ease of maintenance, and low equipment replacement cost.

## SALARIES

In 1996, median weekly earnings of full-time electronic equipment repairers were $619. The middle 50 percent earned between $444 and $802. The bottom 10 percent earned less than $329, while the top 10 percent earned more than $979. Median weekly earnings varied widely by occupation and the type of equipment repaired, as follows:

| | |
|---|---|
| Telephone installers and repairers | $717 |
| Electronic repairers, communications and industrial equipment | 602 |
| Office machine repairers | 582 |
| Data processing equipment repairers | 573 |

Central office installers, central office technicians, PBX installers, and telephone installers and repairers employed by AT&T and the Bell Operating Companies and represented by the Communications Workers of America and the International Brotherhood of Electrical Workers earned between $279 and $962 a week in 1996.

According to a survey of workplaces in 160 metropolitan areas, beginning maintenance electronics technicians had median earnings of $11.50 an hour in 1995, with the middle half earning between $10.50 and $13.25 an hour. The most experienced repairers had median earnings of $20.13 an hour, with the middle half earning between $18.24 and $22.12 an hour.

## RELATED FIELDS

Workers in other occupations who repair and maintain the circuits and mechanical parts of electronic equipment include appliance and power tool repairers, automotive electricians, broadcast technicians, electronic organ technicians, and vending machine repairers. Electronics engineering technicians may also repair electronic equipment as part of their duties.

## INTERVIEW
### Thérèse Heckel
## Switching Equipment Technician

Thérèse Heckel worked as a switching equipment technician and troubleshooter for South Central Bell Telephone Company (now Bell-South) in New Orleans, Louisiana, and Birmingham, Alabama, from 1974 until her retirement in 1997.

*She earned her BA in English from Loyola of the South (Evening Division) in New Orleans in 1974 and also did some graduate work toward an MBA and an MSW.*

*She attended various AT&T/Bell schools for her technical training in New Orleans; Birmingham; San Antonio; Dublin, Ohio; and Lisle, Illinois.*

## How Thérèse Heckel Got Started

"This field was being newly opened up to women, and I was bored with the clerical women's jobs. I wanted a challenge. It didn't hurt that the salary was much higher; we were graduated up to the same pay as the men who previously dominated this job field.

"After taking entrance tests in basic electricity (which I studied on my own as well as at the Union Hall, which provided free classes), I qualified to be trained as a switching equipment technician. After the training was completed, I was on probation, and after a year I was considered qualified.

"At the time, the tradition in any Bell company was that they trained you because they not only wanted the work done their way, but because there was no other phone company. The Bell System had their own classes in electricity, electronics, and any advanced telecommunications repair that needed to be done as a function of our job titles. I began in traditional 'female jobs,' but took various 'male jobs' when they were first opened to us. I began in mechanical crossbar switching in 1974, but as the technology advanced, I progressed to the electronic switching and troubleshooting jobs. But the end result of the job was the same—supplying dial tone and line features to the customer, and troubleshooting the lines when the customer could not use the service."

## What the Job's Really Like

"The job is a twenty-four–hour, seven-days-a-week job, doing behind-the-scenes repair on local and (in the old days) toll lines. My job was inside/network repair, rather than the outside, pole-climbing, customer repair. I was in charge of the dial tone.

"A customer calls to say her line won't work—for example, no dial tone. A ticket is generated and sent to the inside repair technicians, and the line is tested. If the trouble is in the telephone exchange, another ticket is generated to have that technician verify the hardware connected to the line.

"Sometimes it's the line that is bad, and it's fixed individually. At other times, it can be an office problem, and the switching equipment needs testing and correcting.

"When I did this central office job I was working with equipment and circuit packs, and often spoke with the customer to get the line fixed. Now, much of the work is centralized and can be fixed via a computer network that extends throughout the states the Bell company serves.

"Although a forty-hour week is in the contract, there is usually considerable overtime if a major outage or repair crisis occurs. New employees may be requested to take this overtime as it goes on a seniority basis, as does the choice of shift and vacation times.

"I liked troubleshooting and finding and fixing a problem. I also liked the satisfaction of calling the customer and telling her the phone line is back in service (with the proper apologies for the problem).

"What I didn't like was the sedentary nature the job acquired. Prior to computers, we had to use our meters and troubleshooting equipment, drawings, and brains to diagnose problems, and there was more job satisfaction because of the energy expended. We moved around the huge offices and climbed twenty-two-foot ladders and did rewiring and so forth. Now, you sit and type into a computer, and it does a lot of the work. However, I guess most jobs are computer-based in this day and age.

"The salary at top pay is over $800 a week. Top pay is reached with five years of service. New hires would negotiate the pay, per the minimums listed in the CWA union contract."

## Expert Advice

"The desire to see a problem through to the end is crucial. So is being able to deal with people in a friendly, helpful manner!

"Patience and courteous teamwork will get you more help from the various other departments that you will have to work with to see the problem through from start to finish.

"Having a technical or computer background is essential. Even a two-year associates degree will give you an edge. However, there is also an extensive entrance test just to get hired at BellSouth, and your test and interview scores determine whether you will be placed into the title. The current title for the switching equipment technician is electronic technician, or ET."

## INTERVIEW
### Michael John
## Computer Bench Technician

Michael John has been a bench technician with Hendersin Technical Services, a computer repair shop in Cocoa Beach, Florida, since 1996. He earned his AS in computer engineering technology from Keiser College in Melbourne, Florida.

### How Michael John Got Started

"I worked for Office Depot and enjoyed working and messing around with the computers there, so I got my own. I liked it so much that I decided to make a career out of it.

"I went to a technical school, and that gave me most of my training. This is my first computer job. I got it because my teacher also works for the place that I work for now. He thought that I did so well and learned so quickly that he recommended me to the owner."

### What the Job's Really Like

"My duties are almost self-explanatory. I build, repair, and sell computer systems. My typical day starts at 8:30 A.M. I am usually the first one in the shop, so I turn on the prebuilt systems at the front of the store. I make sure that the front is clean and looks good for the day. Next, I start on the repairs that are in the shop from the days before.

"The hands-on job of computer repair is so awesome. I love to get into a machine and just look at how it ticks. The shop that I work for is about five years old, so the business is picking up. I love to stay busy and like having something to do when I get done with the job at hand. I usually work forty hours a week, 8:30 to 5:00, with a half-hour lunch. The work atmosphere is great. There are only four of us, including the owner. We sit and joke and carry on while working—it makes the day go by faster.

"What I like the most about my job is the feeling I get when I figure something out that no one else has been able to. What I like the least is that because I am the one with the most retail training, I have to do the returns and stocking most of the time.

"Most computer technicians make anywhere from $8 to $20 an hour. My salary is on the lower end because I just started in the business with no experience."

## Expert Advice

"The first thing to do is get computer repair training. The best kind is a technical college that will give you hands-on training. Once you finish the course, and I mean no more than a week after, take the A+ exam. This exam certifies you as an expert technician. To do well in this field, you need to be a logical thinker. If you are not, but still enjoy computers, the training can sometimes help."

# FOR MORE INFORMATION

For career and certification information, contact:

> The International Society of Certified Electronics Technicians
> 2708 West Berry Street
> Fort Worth, TX 76109

For certification, career, and placement information, contact:

> Electronics Technicians Association
> 602 North Jackson
> Greencastle, IN 46135

For information on the telephone industry and career opportunities, contact:

United States Telephone Association
1401 H Street NW, Suite 600
Washington, DC 20005-2136

International Brotherhood of Electrical Workers
Telecommunications Department
1125 15th Street NW, Room 807
Washington, DC 20005

For information on electronic equipment repairers in the telephone industry, write to:

Communications Workers of America
Department of Apprenticeships, Benefits, and Employment
501 3rd Street NW
Washington, DC 20001

# CHAPTER 9

# Musical Instrument Repairers and Tuners

**EDUCATION**
H.S. Required

**$$$ SALARY**
$15,000 to $48,800

## OVERVIEW

Musical instruments are a source of entertainment and recreation for millions of people. Maintaining these instruments so they perform properly is the job of musical instrument repairers and tuners. Good hearing, mechanical aptitude, and manual dexterity are necessary to do the job. The occupation includes piano tuners and repairers (often called *piano technicians*); pipe organ tuners and repairers; and brass, woodwind, percussion, or string instrument repairers.

*Piano tuners* adjust piano strings to the proper pitch. A string's pitch is the frequency at which it vibrates and produces sound when it is struck by one of the piano's wooden hammers. Tuners first adjust the pitch of the "A" string. Striking the key, the tuner compares the string's pitch with that of a tuning fork. Using a tuning hammer (also called a *tuning lever* or *wrench*), the tuner turns a steel pin to tighten or loosen the string until its pitch matches that of the tuning fork. The pitch of each of the other strings is set in relation to the "A" string. The standard 88-key piano has 230 strings and can be tuned in about an hour and a half.

A piano has thousands of wooden, steel, iron, ivory, and felt parts that can be plagued by an assortment of problems. It is the task of *piano repairers* to locate and correct these problems. In addition to repair work, piano repairers may also tune pianos.

To diagnose problems, repairers talk with customers before partially dismantling a piano to inspect its parts. Repairers may realign moving parts, replace old or worn ones, or completely rebuild pianos. Repairers use common hand tools as well as special ones, such as regulating, repining, and restringing tools.

Some piano tuners service pianos that have built-in computers that control humidity, assist in recording, or allow the piano to operate as an automatic player piano. Piano repair work will increasingly require some knowledge of electronics, as sales of sophisticated pianos increase and people decide to upgrade their older pianos.

*Pipe organ repairers* tune, repair, and install organs that make music by forcing air through flue pipes or reed pipes. The flue pipe sounds when a current of air strikes a metal lip in the side of the pipe. The reed pipe sounds when a current of air vibrates a brass reed inside the pipe.

To tune an organ, repairers first match the pitch of the "A" pipes with that of a tuning fork. The pitch of other pipes is set by comparing it to that of the "A" pipes. To tune a flue pipe, repairers move the metal slide that increases or decreases the pipe's "speaking length." To tune a reed pipe, the tuner alters the length of the brass reed. Most organs have hundreds of pipes, so often a day or more is needed to completely tune an organ.

Pipe organ repairers locate problems, repair or replace worn parts, and clean pipes. Repairers also assemble organs on site in churches and auditoriums, following manufacturer's blueprints. They use hand and power tools to install and connect the air chest, blowers, air ducts, pipes, and other components. They may work in teams or be assisted by helpers. Depending on the size of the organ, a job may take from several weeks to several months.

*Violin repairers* adjust and repair bowed instruments, such as violins, violas, and cellos, using a variety of hand tools. They find defects by inspecting and playing instruments. They remove cracked or broken sections, repair or replace defective parts, and restring instruments. They also fill in scratches with putty, sand rough spots, and apply paint or varnish.

*Guitar repairers* inspect and play the instrument to determine defects. They replace levels using hand tools, and fit wood or metal parts. They reassemble and string guitars.

Brass and woodwind instruments include trumpets, cornets, French horns, trombones, tubas, clarinets, flutes, saxophones, oboes, and bassoons. *Brass and wind instrument repairers* clean, adjust, and repair these instruments. They move mechanical parts or play scales to find defects. They may unscrew and remove rod pins, keys, and pistons, and remove soldered parts using gas torches. They repair dents in metal instruments using mallets or burnishing tools. They fill cracks in wood instruments by inserting pinning wire and covering them with filler. Repairers also inspect instrument keys and replace worn pads and corks.

*Percussion instrument repairers* work on drums, cymbals, and xylophones. In order to repair a drum, they remove drum tension rod screws and rods by hand or by using a drum key. They cut new drumheads from animal skin, stretch the skin over rimhoops, and tuck it around and under the hoop using hand tucking tools. To prevent a crack in a cymbal, gong, or similar instrument from advancing, repairers may operate a drill press or hand power drill to drill holes at the inside edge of the crack. Another technique they may use involves cutting out sections around the cracks using shears or grinding wheels. They also replace the bars and wheels of xylophones.

Although they may suffer small cuts and bruises, the work of musical instrument repairers and tuners is relatively safe. Most brass, woodwind, percussion, and string instrument repairers work in repair shops or music stores. Piano and organ repairers and tuners usually work on instruments in homes, schools, and churches and may spend several hours a day driving. Salaried repairers and tuners work out of a shop or store; the self-employed generally work out of their homes.

# TRAINING

For musical instrument repairer and tuner jobs, employers prefer people with postsecondary training in music repair technology.

Some musical instrument repairers and tuners learn their trade on the job as apprentices or assistants, but employers willing to provide on-the-job training are difficult to find. A few music stores, large repair shops, and self-employed repairers and tuners hire inexperienced people as trainees to learn how to tune and repair instruments under the supervision of experienced workers. Trainees may sell instruments, clean up, and do other routine work. Usually two to five years of training and practice are needed to become fully qualified.

A small number of technical schools and colleges offer courses in piano technology or brass, woodwind, string, and electronic musical instrument repair. A few music repair schools offer one- or two-year courses. There are also home study (correspondence school) courses in piano technology. Graduates of these courses generally refine their skills by working for a time with an experienced tuner or technician.

Music courses help develop the student's ear for tonal quality. The ability to play an instrument is helpful. Knowledge of woodworking is useful for repairing instruments made of wood.

Repairers and tuners need good hearing, mechanical aptitude, and manual dexterity. For those dealing directly with customers, a neat appearance and a pleasant, cooperative manner are important.

Musical instrument repairers keep up with developments in their fields by studying trade magazines and manufacturers' service manuals. The Piano Technicians Guild helps its members improve their skills through training conducted at local chapter meetings and at regional and national seminars. Guild members also can take a series of tests to earn the title *registered piano technician*. The National Association of Professional Band Instrument Repair Technicians offers similar programs and scholarships, and a trade publication. Its members specialize in the repair of woodwind, brass, string, and percussion instruments. Repairers and technicians who work for large dealers, repair shops, or manufacturers can advance to supervisory positions or go into business for themselves.

# JOB OUTLOOK

Musical instrument repairers and tuners held about 9,000 jobs in 1996. Most technicians work on pianos. About half are self-employed. About eight of ten wage and salary repairers and tuners work in music stores, and most of the rest work in repair shops or for musical instrument manufacturers.

Musical instrument repairer and tuner jobs are expected to increase more slowly than the average for all occupations through 2006. Replacement needs will provide the most job opportunities as many repairers and tuners near retirement age. The small number of openings, due to both growth and replacement needs, is very low relative to other occupations. Because training is difficult to receive—only a few schools offer training programs, and few experienced workers are willing to take on apprentices—opportunities should be excellent for those who do receive training.

Several factors are expected to influence the demand for musical instrument repairers and tuners. The number of people employed as musicians will increase, mainly due to a slight increase in the number of students of all ages playing musical instruments. Because instruments are quite expensive to purchase, growing numbers of instrument repairers will be needed to work on rental equipment leased to students, schools, and other organizations.

# SALARIES

Musical instrument repairers and tuners who usually worked full time had median weekly earnings of $559 in 1996. The middle 50 percent earned between $428 and $855 a week. The lowest 10 percent earned less than $300, while the highest 10 percent earned more than $939 a week. Earnings were generally higher in urban areas.

# RELATED FIELDS

Musical instrument repairers need mechanical aptitude and good manual dexterity. Electronic home entertainment equipment repairers, vending machine servicers and repairers, home appliance and power tool repairers, and computer and office machine repairers all require similar talents.

# INTERVIEW
## Jim Foreman
## Musical Instrument Repairer

*Jim Foreman worked for ten years as a professional musician, music teacher, musical instrument repairman, and piano mover in the Los Angeles area.*

## How Jim Foreman Got Started

"I was a musician in my heart of hearts. Most musical instrumentalists develop a set of rudimentary repair skills for their instrument of choice. As a guitarist and bass player, I found it necessary to perform such tasks as changing strings, filing and seating frets, filing and honing the nut and the bridge of the guitar, changing the machine heads, oiling and adjusting the machine heads, soldering connections for electric guitars, replacing the pick-ups for electric guitars, and so forth. Thus, by just simply being a guitar player, I had developed a set of skills that I could sell under the right circumstances.

"The right circumstances appeared during my college days, when I found myself employed as a guitar teacher in a music store. The store sold many guitars, and I discovered several of those instruments were soon in need of repair. In addition, some of the guitar players in the area preferred that I perform routine maintenance on their 'axes' (musician slang). Thus, armed with a glue gun and the ability to spot a sprung bridge or the seventh fret causing that oh-so-annoying string buzz, I set forth in a world in need of repair.

"My first training began at the age of thirteen years in junior high school wood shop and metal shop. I took these classes before I got into music and I say now that this sequence is better than the other way around.

"Around the age of fourteen, I began to study music. Like many of the baby boom generation, I related to music and the culture of the sixties as my artistic pabulum. Music quickly became my focal point in life. All things musical interested me, and this interest evolved into repair skills.

"I was very lucky to have observed repair procedures such as the placement of clamps, usage of different types of sand paper, and the use of diverse glues and their drying times. I worked in a shop where all manner of musical instruments were repaired (school band instruments, orchestra instruments, and pop music instruments). I had the opportunity to consult other repairers, who often offered a unique approach to a difficult repair problem."

## What the Job's Really Like

"As an independent contractor, I didn't have a boss hanging over me, making demands, or seeing to it that I punched my time card. I did have to satisfy every customer, usually within one to two weeks of the acceptance of the job. My philosophy was to make the customer happy. This involved many skills, including psychology as well as technical skills.

"With a flexible schedule, I could come in during the middle of the night if I needed or wanted to, or I could keep regular hours. I found it necessary to keep both irregular and regular hours. In order to repair some of the more difficult problems, I needed time to dream, converse with other repairers, and research. Sometimes, the hours were light; there were other times when I struggled to keep up with the volume of instruments in need of repair.

"Since I worked in a medium-sized music store, there was always something that needed repair. I would take any job, from the little kid with a $20 guitar that had come apart on him, to the highly regarded recording artist who wanted me to design and build an instrument that existed only in his imagination.

"Occasionally, a very exotic job would come my way (such as custom-building instruments). On those jobs, I would spend as much time as possible developing an instrument that I would be happy with. For example, an internationally renowned musical artist, L. Subramaniam, asked me to do work for him. This gentleman petitioned me to design and build what was then a nonexistent violin for his exclusive use.

"As in all situations involving self-regulated employment, I loved the autonomy. Ultimately, I did have to deal with a person who had the power to fire me—the owner of the music store itself. But I felt that this person dealt fairly with me. He saw to it that if an item was needed in the shop, it was purchased quickly and well kept. For example, one of the most useful power tools for repairing musical instrument is the Dremel Moto-Tool (now called the Dremel Multi-Pro Tool). I remember that I discovered the value of this appliance by borrowing one from another repairer. Suffice it to say that I was knocked out by that utensil, and I had to have one. Well, in those days the Dremel cost close to $100. As a starving artist, I definitely could not afford to spend that much money on a tool. So I worked up some figures on how much time I would save by having the device, and added some other general information such as how the quality of my work would improve with the apparatus. The owner listened carefully, asked me to show him two or three examples of how the tool performed miracles, and quickly bought the gizmo for the shop. I managed to double the number of repairs passing through the owner's cash register after he purchased the gadget, and believe me, he noticed that!

"There are downsides to musical instrument repair work from time to time. Some people are what I call musically challenged. Musically challenged people do not have a sense of pitch that corresponds with any tonal system that I am aware of. I simply cannot teach music to them or work on their instruments because when I try to, I begin to lose my own sense of rhythm.

"Another one of the downsides of being a musical instrument repairer was that I could not refuse to work for certain people. This included parents who buy musical instruments for children who are not the least bit interested in learning and playing music. The poor guitar, bass, mandolin, violin, banjo, or whatever that is put in the hands of such a progeny soon shows signs of being abused, trashed, and otherwise mistreated.

"Sadly, the art form of acoustic guitar repair and the artists who make acoustic guitars (luthiers) are suffering from a lack of available woods such as African ebony and Brazilian rosewood. Many of the modern luthiers of electric guitars are now making their guitars from composite materials such as plastics and fiberglass. This is because you can make an electric guitar out of just about anything, from an ironing board to a telephone book. With electric guitars, the electronics are more important than what the guitar's body is made of."

## Expert Advice

"When you have even a general idea as to what you are going to charge for the services you render, make a large and easy-to-read sign setting forth the prices, and post it in a place where it will be seen.

"If you want to get paid by the hour, that's fine. You should be aware, however, that most musical instrument repairers are independent contractors. It is my opinion that when a skilled repairer works for an hourly rate, he or she is getting paid less than the independent contractor.

"I have seen some repairers make as much as $60,000 a year. These were the top-drawer repairers who worked in the crème-de-la-crème music stores and worked for professional musicians of the variety found in big-time L.A. recording studios, or in celebrity bands. In my job as a part-timer working his way through college and looking to make just enough money and the most money possible, I made considerably less than that.

"But income aside, I am of the opinion that a repairer should be able to play the instrument he or she is repairing. Not all repairers are good players, and some excellent repairers can't play a lick. But I do not understand how these people find out what the instrument's problem is or could be. So, it is not absolutely necessary for a repairer to play, but I find that it sure helps.

"An aspiring musical instrument repairer should buy his or her own instrument and start performing tasks upon that instrument. Also, and maybe more importantly, the newbie should find what I call a 'master' and do whatever it takes to learn everything the master knows, if possible. Offer to apprentice or be the intern for the shop. You usually won't get paid in these classifications, but you are getting free training."

# FOR MORE INFORMATION

Details about job opportunities may be available from local music instrument dealers and repair shops.

For general information about piano technicians and a list of schools offering courses in piano technology, write to:

Piano Technicians Guild
3930 Washington Street
Kansas City, MO 64111-2963

For general information on musical instrument repair, write to:

National Association of Professional Band Instrument
   Repair Technicians (NAPBIRT)
P.O. Box 51
Normal, IL 61761

# CHAPTER 10
# Heating, Air Conditioning, and Refrigeration Technicians

**EDUCATION**
H. S. Required

**SALARY**
$14,000 to $46,000

## OVERVIEW

What would those living in Chicago do without heating, those in Miami do without air conditioning, or blood banks in all parts of the country do without refrigeration? Heating and air conditioning systems control the temperature, humidity, and the total air quality in residential, commercial, industrial, and other buildings. Refrigeration systems make it possible to store and transport food, medicine, and other perishable items. Heating, air conditioning, and refrigeration technicians install, maintain, and repair such systems.

Heating, air conditioning, and refrigeration systems consist of many mechanical, electrical, and electronic components, including motors, compressors, pumps, fans, ducts, pipes, thermostats, and switches. In central heating systems, for example, a furnace heats air that is distributed throughout the building via a system of metal or fiberglass ducts. Technicians must be able to maintain, diagnose, and correct problems throughout the entire system. To do this, they may adjust system controls to recommended settings and test the performance of the entire system using special tools and test equipment.

Although they are trained to do both, technicians generally specialize in either installation or maintenance and repair. Some further specialize in one type of equipment—for example, oil burners, solar panels, or commercial refrigerators. Technicians may work for large or small contracting companies or directly for a manufacturer or wholesaler. Those working for smaller operations tend to do both installation and servicing, and work with heating, cooling, and refrigeration equipment.

Furnace installers, also called *heating equipment technicians*, follow blueprints or other specifications to install oil, gas, electric, solid-fuel, and multiple-fuel heating systems. After putting the equipment in place, they install fuel and water supply lines, air ducts and vents, pumps, and other components. They may connect electrical wiring and controls and check the unit for proper operation. To ensure the proper functioning of the system, furnace installers often use combustion test equipment such as carbon dioxide and oxygen testers.

After a furnace has been installed, technicians often perform routine maintenance and repair work in order to keep the system operating efficiently. During the fall and winter, for example, when the system is used most, they service and adjust burners and blowers. If the system is not operating properly, they check the thermostat, burner nozzles, controls, or other parts in order to diagnose and then correct the problem. During the summer, when the heating system is not being used, technicians do maintenance work, such as replacing filters and vacuum-cleaning vents, ducts, and other parts of the system that may accumulate dust and impurities during the operating season.

Air conditioning and refrigeration technicians install and service central air conditioning systems and a variety of refrigeration equipment. Technicians follow blueprints, design specifications, and manufacturers' instructions to install motors, compressors, condensing units, evaporators, piping, and other components. They connect this equipment to the duct work, refrigerant lines, and electrical power source. After making the connections, they charge the system with refrigerant, check it for proper operation, and program control systems.

When air conditioning and refrigeration equipment breaks down, technicians diagnose the problem and make repairs. To do this, they may test parts such as compressors, relays, and thermostats. During the winter, air conditioning technicians inspect the systems and do required maintenance, such as overhauling compressors.

When servicing equipment, heating, air conditioning, and refrigeration technicians must use care to conserve, recover, and recycle chlorofluorocarbon (CFC) and hydrochlorofluorocarbon (HCFC) refrigerants used in air conditioning and refrigeration systems. The release of CFCs and HCFCs is thought to contribute to the depletion of the stratospheric ozone layer, which protects plant and animal life from ultraviolet radiation. Technicians conserve the refrigerant by making sure that there are no leaks in the system; they recover it by venting the refrigerant into proper cylinders; and they recycle it for reuse with special filter-dryers.

Heating, air conditioning, and refrigeration technicians use a variety of tools, including hammers, wrenches, metal snips, electric drills, pipe cutters and benders, measurement gauges, and acetylene torches, to work with refrigerant lines and air ducts. They use voltmeters, thermometers, pressure gauges, manometers, and other testing devices to check air flow, refrigerant pressure, electrical circuits, burners, and other components.

Cooling and heating systems sometimes are installed or repaired by other craft workers. For example, on a large air conditioning installation job, especially where workers are covered by union contracts, duct work might be done by sheet-metal workers; electrical work by electricians; and installation of piping, condensers, and other components by plumbers and pipefitters. Room air conditioners and household refrigerators usually are serviced by home appliance repairers.

Heating, air conditioning, and refrigeration technicians work in homes, supermarkets, hospitals, office buildings, factories—anywhere there is climate control equipment. They may be assigned to specific job sites at the beginning of each day, or if they are making service calls, they may be dispatched to jobs by radio or telephone.

Technicians may work outside in cold or hot weather or in buildings that are uncomfortable because the air conditioning or heating equipment is broken. In addition, technicians often work in awkward or cramped positions and sometimes are required to work in high places. Hazards include electrical shock, burns, muscle strains, and other injuries from handling heavy equipment. Appropriate safety equipment is necessary when handling refrigerants since contact can cause skin damage, frostbite, or blindness. Inhalation of refrigerants when working in confined spaces is also a possible hazard, and may cause asphyxiation.

Technicians usually work a forty-hour week, but during peak seasons they often work overtime or irregular hours. Maintenance workers, including those who provide maintenance services under contract, often work evening or weekend shifts, and are on call. Most employers try to provide a full work week year-round by doing both installation and maintenance work, and many manufacturers and contractors now provide or even require service contracts. In most shops that service both heating and air conditioning equipment, employment is very stable throughout the year.

# TRAINING

Because of the increasing sophistication of heating, air conditioning, and refrigeration systems, employers prefer to hire those with technical school or apprenticeship training. A sizable number of technicians, however, still learn the trade informally on the job.

Many secondary and postsecondary technical and trade schools, junior and community colleges, and the armed forces offer six-month to two-year programs in heating, air conditioning, and refrigeration. Students study theory, design, and equipment construction, as well as electronics. They also learn the basics of installation, maintenance, and repair.

Apprenticeship programs are frequently run by joint committees representing local chapters of the Air-Conditioning Contractors of America, the Mechanical Contractors Association of America, the National Association of Plumbing-Heating-Cooling Contractors, and locals of the Sheet Metal Workers' International Association or the United Association of Journeymen and Apprentices of the Plumbing and Pipefitting Industry of the United States and Canada. Other apprenticeship programs are sponsored by local chapters of the Associated Builders and Contractors and the National Association of Home Builders.

Formal apprenticeship programs generally last three or four years and combine on-the-job training with classroom instruction. Classes cover such subjects as the use and care of tools, safety practices, blueprint reading, and air conditioning theory. Applicants for these programs must have a high school diploma or equivalent.

Those who acquire their skills on the job usually begin by assisting experienced technicians. They may begin performing simple tasks such as carrying materials, insulating refrigerant lines, or cleaning furnaces. In time, they move on to more difficult tasks, such as cutting and soldering pipes and sheet metal and checking electrical and electronic circuits.

Courses in shop math, mechanical drawing, applied physics and chemistry, electronics, blueprint reading, and computer applications provide a good background for those interested in entering this occupation. Some knowledge of plumbing or electrical work is also helpful. A basic understanding of microelectronics is becoming more important because of the increasing use of this technology in solid-state equipment controls. Because technicians frequently deal directly with the public, they should be courteous and tactful, especially when dealing with an aggravated customer. They also should be in good physical condition because they sometimes have to lift and move heavy equipment.

All technicians who purchase or work with refrigerants must be certified so that they know how to handle them properly. To become certified to purchase and handle refrigerants, technicians must pass a written examination specific to the type of work in which they specialize.

The three possible areas of certification are:

- Type I—servicing small appliances

- Type II—high-pressure refrigerants

- Type III—low-pressure refrigerants

Exams are administered by organizations approved by the Environmental Protection Agency, such as trade schools, unions, contractor associations, or building groups. Though no formal training is required for certification, training programs designed to prepare workers for the certification examination, as well as for general skills improvement training, are provided by heating and air conditioning equipment manufacturers; the

Refrigeration Service Engineers Society (RSES); the Air Conditioning Contractors of America (ACCA); the Mechanical Service Contractors of America; local chapters of the National Association of Plumbing-Heating-Cooling Contractors; and the United Association of Plumbers and Pipefitters. RSES, along with some other organizations, also offers basic self-study courses for individuals with limited experience.

In addition to understanding how systems work, technicians must be knowledgeable about refrigerant products and legislation and regulation that govern their use. Advancement usually takes the form of higher wages. Some technicians, however, may advance to positions such as supervisor or service manager. Others may move into areas such as sales and marketing. Those with sufficient money and managerial skill can open their own contracting businesses.

# JOB OUTLOOK

Heating, air conditioning, and refrigeration technicians held about 233,000 jobs in 1996. More than one-half of these work for cooling and heating contractors. The remainder are employed in a wide variety of industries throughout the country, reflecting a widespread dependence on climate control systems.

Some work for fuel oil dealers, refrigeration and air conditioning service and repair shops, and schools. Others are employed by the federal government, hospitals, office buildings, and other organizations that operate large air conditioning, refrigeration, or heating systems. Approximately one of every eight technicians is self-employed.

Job prospects for highly skilled air conditioning, heating, and refrigeration technicians are expected to be very good, particularly for those with technical school or formal apprenticeship training to install, remodel, and service new and existing systems. In addition to job openings created by employment growth, thousands of openings will result from the need to replace workers who transfer to other occupations or leave the labor force.

Employment of heating, air conditioning, and refrigeration technicians is expected to increase about as quickly as the average for all occupations through 2006. As the population and economy grow, so does the demand for new residential, commercial, and industrial climate control systems. Technicians who specialize in installation work may experience periods of unemployment when the level of new construction activity declines, but maintenance and repair work usually remains relatively stable. People and businesses depend on their climate control systems and must keep them in good working order, regardless of economic conditions.

Concern for the environment and energy conservation should continue to prompt the development of new energy-saving heating and air-conditioning systems. An emphasis on better energy management should lead to the replacement of older systems and the installation of newer, more efficient systems in existing homes and buildings. Also, demand for maintenance and service work should increase as businesses and home owners strive to keep systems operating at peak efficiency. Regulations prohibiting the discharge of CFC and HCFC refrigerants and banning CFC production by 2000 also should continue to result in demand for technicians to replace many existing systems, or modify them to use new environmentally safe refrigerants. In addition, the continuing focus on improving indoor air quality should contribute to the growth of jobs for heating, air conditioning, and refrigeration technicians. Also, certain businesses contribute to a growing need for refrigeration. For example, nearly 50 percent of products sold in convenience stores require some sort of refrigeration. Supermarkets and convenience stores have a very large inventory of refrigerated equipment. This huge inventory will also create increasing demand for service technicians in installation, maintenance, and repair.

# SALARIES

Median weekly earnings of air conditioning, heating, and refrigeration technicians who worked full time were $536 in 1996. The middle 50 percent earned between $381 and $701. The lowest 10 percent earned less than $287 a week, and the top 10 percent earned more than $887 a week.

Apprentices usually begin at about 50 percent of the wage rate paid to experienced workers. As they gain experience and improve their skills, they receive periodic increases until they reach the wage rate of experienced workers.

Heating, air conditioning, and refrigeration technicians enjoy a variety of employer-sponsored benefits. In addition to typical benefits like health insurance and pension plans, some employers pay for work-related training and provide uniforms, company vans, and tools.

Nearly one out of every six heating, air conditioning, and refrigeration technicians is a member of a union. The unions to which the greatest numbers of technicians belong are the Sheet Metal Workers' International Association and the United Association of Journeymen and Apprentices of the Plumbing and Pipefitting Industry of the United States and Canada.

## RELATED FIELDS

Heating, air conditioning, and refrigeration technicians work with sheet metal and piping, and repair machinery, such as electrical motors, compressors, and burners. Other workers who have similar skills are boilermakers, electrical appliance servicers, electricians, plumbers and pipefitters, sheet-metal workers, and duct installers.

## INTERVIEW
### Troy Scott McClure
### Owner of a Refrigeration, Heating, and Air Conditioning Business

Troy Scott McClure owns his own refrigeration, heating, and air conditioning business in Iowa Park, Texas. He earned his BS in education in 1996 from Midwestern State University in Wichita Falls, Texas. He has been in the HVAC field since 1983.

## How Troy Scott McClure Got Started

"I had a terrible job at a car dealership, and an opportunity came up in the air conditioning (AC) business. Basically, I learned as I went. My boss showed me some things, but his philosophy was that you learned best by doing. Now I have attended a few seminars put on by manufacturers of AC equipment.

"Timing is everything in life; things just worked out for me. My father was working with someone in 1983 who had an air conditioning business on the side. He was short on help one day and casually mentioned it around the dinner table at the firehouse where my father worked. The rest, of course, is history.

"After a while I became fed up working for someone else and making him lots of money, so I ventured out on my own."

## What the Job's Really Like

"My duties are installing AC equipment, servicing, sending bills, collecting bills, and other boring book work. I am the busiest when the weather is either really hot or really cold.

"There is no such thing as a typical day in my business. This summer has been really hot, so usually I'll get up at daybreak to get any attic work done, and then I move on to service work. Usually I eat on the go. My personal record for service calls finished in one day is ten. Other days are spent installing equipment. A total system change can take a whole day, and the challenge is to make service calls too.

"It can go from relaxed to swamped in minutes. Usually the fall and spring months are more relaxed, with the busiest times being in the summer.

"The work is interesting at times. Some customers can be a blast, while others can be a pain. But, being self-employed, I am selective in whom I work for. There is a lot of driving between jobs, which can be a drag.

"During summer months, I can work seventy hours if my body allows, while in the mild weather I may not work thirty hours a week. Of course, being self-employed is great. The more I work, the more money I make. I don't have the desire to become a huge company, so mainly I do what I can do. Right now my retired fireman father helps me occasionally when I need assistance. I enjoy meeting new customers and taking care of my old customers' needs. I tend to develop a relationship with my customers and, as strange as this might sound, also with their equipment.

"I like being outside during most of the work, but the downside is attic work and having to get under houses, which is the least desirable of my duties. The most desirable is collecting my money after the job is done. I always worry about getting stiffed on a job.

"The upside is that you can make good money! I charge $36 per hour. The good money is when you can sell equipment, though. Also, on parts you get the markup. You get paid for the labor as well as the markup on the equipment.

"Duct runners start out at the lowest, around $6 an hour. New service techs make about $10 an hour. This area of the country is one of the lowest-paying of anywhere, though."

## Expert Advice

"Bottom line—this business is not for wimps. If you can't stand the heat, you gotta get out of the kitchen. This trade and the others give you some of the few opportunities to become your own boss someday. I worked for my old boss for five years and then went on my own. I learned as I went, basically, but there are trade schools for the profession. The key is to find a good boss who will give you the freedom to be your own person. There are lots of awful people to work for, but I was fortunate to get to work for a great man who helped me on my road to self-employment.

"One more thing. There are lots of bad guys out there in the trades who are very dishonest and try to get as much money from their customers as possible. I despise these guys. I go home every night feeling good about taking care of my people and charging them a fair price."

# FOR MORE INFORMATION

For more information about employment and training opportunities in this trade, contact local vocational and technical schools; local heating, air conditioning, and refrigeration contractors; a local of the unions previously mentioned; a local joint union-management apprenticeship committee; a local chapter of the Associated Builders and Contractors; or the nearest office of the state employment service or state apprenticeship agency.

For information on career opportunities and training, write to:

Air Conditioning and Refrigeration Institute
4301 North Fairfax Drive, Suite 425
Arlington, VA 22203

Air Conditioning Contractors of America
1712 New Hampshire Avenue, NW
Washington, DC 20009

Associated Builders and Contractors
1300 North 17th Street
Rosslyn, VA 22209

Refrigeration Service Engineers Society
1666 Rand Road
Des Plaines, IL 60016-3552

Home Builders Institute
National Association of Home Builders
1201 15th Street, NW
Washington, DC 20005

Mechanical Service Contractors of America
1385 Piccard Drive
Rockville, MD 20850-4329

National Association of Plumbing-Heating-Cooling
   Contractors
180 S. Washington Street
P.O. Box 6808
Falls Church, VA 22046

New England Fuel Institute
P.O. Box 9137
Watertown, MA 02272

# CHAPTER 11

# Vending Machine Repairers

### 🎓 EDUCATION
H. S. Required

### 💲💲💲 SALARY
$10,000 to $35,000

## OVERVIEW

Coin-operated vending machines are a familiar sight. These machines dispense many types of refreshments, from cold soft drinks to hot meals. Vending machine servicers and repairers install, service, and stock these machines and keep them in good working order.

Vending machine servicers periodically visit coin-operated machines that dispense soft drinks, candy, snacks, and other food items. They collect coins from the machines, restock merchandise, change labels to indicate new selections, and adjust temperature gauges so that items are kept at the right temperature. They are also responsible for keeping the machines clean. Because many vending machines dispense food, these workers must comply with state and local public health and sanitation standards.

Servicers make sure machines operate correctly. When checking complicated electrical and electronic machines, such as beverage dispensers, they make sure that the machines mix drinks properly and that refrigeration and heating units work correctly. On the relatively simple gravity-operated machines, servicers check handles, springs, plungers, and merchandise chutes. They also test coin and change-making mechanisms.

When installing the machines, they make the necessary water and electrical connections and recheck the machines for proper operation. They also make sure installations comply with local plumbing and electrical codes.

Preventive maintenance—avoiding trouble before it starts—is a major job of these workers. For example, they periodically clean refrigeration condensers, lubricate mechanical parts, and adjust machines to perform properly.

If a machine breaks down, vending machine repairers inspect it for obvious problems, such as loose electrical wires, malfunctions of the coin mechanism, and leaks. If the problem cannot be readily located, they refer to technical manuals and wiring diagrams and use testing devices such as electrical circuit testers to find defective parts. Repairers sometimes fix faulty parts at the site, but they often install replacements and take broken parts to the company shop for repair. When servicing electronic machines, repairers may only have to replace a circuit board or other component. They also repair microwave ovens used to heat food dispensed from machines.

In repair and maintenance work, repairers use hammers, pliers, pipe cutters, soldering guns, wrenches, screwdrivers, and electronic testing devices. In the repair shop, they use power tools, such as grinding wheels, saws, and drills, as well as voltmeters, ohmmeters, oscilloscopes, and other testing equipment.

Vending machine servicers and repairers employed by small companies may both fill and fix machines on a regular basis. These combination servicers-repairers stock machines, collect money, fill coin and currency changers, and repair machines when necessary.

Servicers and repairers also do some clerical work, such as filing reports, preparing repair cost estimates, ordering parts, and keeping daily records of merchandise distributed. However, many of the new computerized machines reduce the paperwork that a servicer performs.

Some vending machine repairers work primarily in company repair shops, but many servicers and repairers spend much of their time on the road visiting machines wherever they have been placed. Vending machines operate around the clock, so repairers often work at night and on weekends and holidays.

Vending machine repair shops generally are quiet, well lit, and have adequate work space. However, when servicing machines on location, the work may be done where pedestrian traffic is heavy, such as in busy supermarkets, industrial complexes, offices, or schools. Repair work is relatively safe, although servicers and repairers must take care to avoid hazards such as electrical shocks and cuts from sharp tools and metal objects. They also must follow safe work procedures, especially when moving heavy vending machines or working with electricity and radiation from microwave ovens.

# TRAINING

Employers generally prefer to hire high school graduates and to train them to fill and fix machines informally on the job by observing, working with, and receiving instruction from experienced repairers. High school or vocational school courses in electricity, refrigeration, and machine repair are an advantage in qualifying for entry-level jobs. Employers usually require applicants to demonstrate mechanical ability, either through their work experience or scoring well on mechanical aptitude tests. Because vending machine servicers and repairers sometimes handle thousands of dollars in merchandise and cash, employers hire persons who have a record of honesty and respect for the law. The ability to deal tactfully with people also is important. A commercial driver's license and a good driving record are essential for most vending machine repairer jobs.

The use of electronics is becoming more prevalent in vending machines, so employers increasingly prefer applicants to have some training in electronics. Technologically advanced machines with features such as multilevel pricing, inventory control, and scrolling messages extensively use electronics and microchip computers. Some vocational high schools and junior colleges offer one- to two-year training programs in basic electronics for vending machine servicers and repairers.

Beginners may start their training with simple jobs such as cleaning or painting machines. They then may learn to rebuild machines—removing defective parts, repairing, adjusting, and testing the machines. Next, they accompany an experienced repairer on service calls, and finally make visits on their own. This learning process may take from six months to three years, depending on the individual's abilities, previous education, types of machines, and the quality of instruction.

The National Automatic Merchandising Association has established an apprenticeship program for vending machine repairers. Apprentices receive 144 hours of home study instruction in subjects such as basic electricity and electronics, blueprint reading, customer relations, and safety. Upon completion of the program, performance and written tests must be passed to become certified.

To learn about new machines, repairers and servicers sometimes attend training sessions sponsored by manufacturers, which may last from a few days to several weeks. Both trainees and experienced workers sometimes take evening courses in basic electricity, electronics, microwave ovens, refrigeration, and other related subjects. Skilled servicers and repairers may be promoted to supervisory jobs.

## JOB OUTLOOK

Vending machine servicers and repairers held about 21,000 jobs in 1996. Most repairers work for vending companies that sell food and other items through machines. Others work for soft drink bottling companies that have their own coin-operated machines. Some work for companies that also own video games, pinball machines, jukeboxes, and similar types of amusement equipment. Although vending machine servicers and repairers are employed throughout the country, most are located in areas with large populations and many coin and vending machines.

Employment of vending machine servicers and repairers is expected to decline through 2006 because improved technology will require servicers and repairers to check on machines less frequently. More vending machines are likely to be installed in industrial plants, hospitals, stores, and other business establishments to meet the public demand for vending machine items. In addition, the range of products dispensed by machine can be expected to increase as vending machines continue to become more automated and more are built with microwave ovens, mini-refrigerators, and freezers. These new machines will need to be repaired and restocked less often, and will contain computers that record sales and inventory data, reducing time-consuming paperwork now done by servicers. Some new machines will use wireless data transmitters to signal the vending machine company when they need to be restocked or repaired. This allows servicers and repairers to be dispatched only when needed, instead of having to check each machine on a regular schedule.

Although employment is expected to decline, there will be job openings as experienced workers transfer to other occupations or leave the labor force. People with some background in electronics should have good job prospects because electronic circuitry is an important component of vending machines. If firms cannot find trained or experienced workers for these jobs, they are likely to train qualified route drivers or hire inexperienced people who have acquired some mechanical, electrical, or electronic training by taking high school or vocational courses.

## SALARIES

According to a survey conducted by the National Automatic Merchandising Association, the average hourly wage rate for nonunion vending machine servicers was $8.66 in 1996. Rates ranged from just under $5 to nearly $17 an hour, depending on the size of the firm and the region of the country. Nonunion repairers averaged $10.38 an hour in 1996, but rates also ranged from about $5.00 to $17.00. Servicers and repairers who were members of unions usually earned slightly more.

Most vending machine repairers work eight hours a day, five days a week, and receive premium pay for overtime. Some union contracts stipulate higher pay for night work and for emergency repair jobs on weekends and holidays.

Some vending machine repairers and servicers are members of the International Brotherhood of Teamsters.

# RELATED FIELDS

Other workers who repair equipment with electrical and electronic components include home appliance and power tool repairers, electronic equipment repairers, and general maintenance mechanics.

## INTERVIEW
### Thomas Walker
### Field Maintenance Worker

Thomas Walker has been in the vending industry since 1981 and has vast experience with repairs and service. He currently does field maintenance for Tampa Bay Vending, Inc., a full-line vending firm in Tampa, Florida.

He earned his AA in management at Wenatchee Community College in Wenatchee, Washington, and has taken numerous courses in vending maintenance.

## How Thomas Walker Got Started

"I was looking for a new career after the army. My father-in-law was in the vending machine repair field, and he loved the industry. I applied at the company's headquarters in person. The position I was offered looked challenging.

"My training was mostly on the job, and I studied at home with journals and manuals and some trial and error."

## What the Job Is Like

"The job is fast-paced and detailed. You have drive time, so you are in the sun and weather, not behind a desk. You work to repair the vending machines that others rely on for food and snacks. Each call is different and needs to be completed quickly, or we could lose a happy customer.

"My duties are:

"One: Respond to service calls.

"Two: Repair all types of vending machines—coffee (ground, instant, fresh-brewed); snack; can drinks; cup drinks; microwaves; food machines; popcorn; speciality vends.

"Three: Perform preventive maintenance—try to fix an item before it is a major repair job.

"Four: Respond to the customers' complaints. They are sometimes very nice, and other times not so pleasant.

"A typical day starts at approximately 6:30 A.M., but what will happen for the next eleven or twelve hours is always unknown. Between 6:30 and 6:45, I start the truck and turn on the company radio. Calls will start coming in from route drivers, and I'll respond to those calls. Then, around 8:30, when our customers begin their day, they'll call in with problems or complaints. Now we are busy!

"These service calls differ in nature, but are usually about a machine not working correctly. There is a time limit to arrive, determine the problem, repair the item, notify the customer of the final outcome, and head for the next call. The delays happen with major repairs such as compressors or tank repairs. These can really back up the calls.

"The job provides an interesting challenge with each service call. Every call you get comes with its own set of problems that need to be solved. Some of these problems you may have already encountered before and can repair quickly. But many times a similar type of problem will have a new kink and will offer you a new challenge, both mentally and physically.

"The people you interact with are as different as the machines. You get to see the city, and on occasion, a movie star here or there. I have met two TV actors and one country singer while working on vending machines.

"The people you face day to day are unique as well. You may deal with a representative you greatly admire, or one you hope to never see again. But I like my main supervisors and the owner of the firm.

"The part I least like about the job is the typical corporate office issues and some of the stress that the time limits can place on the day. I hate the city traffic, but the last I looked, country animals don't have need for vending machines. So the city is the major user of the service I provide, and with it comes the traffic."

## Expert Advice

"Anyone entering the field today will need to be a self-starter who enjoys interacting with people. An outgoing personality helps.

"And you need to enjoy the problem-solving aspects of repair. Take any classes you can find offered on electronics. Dealing with seven or eight different types of machines from as many different manufacturers requires study and a good memory. You also need a basic knowledge of electronics, wiring, diagrams, plumbing, refrigeration, and a solid problem-solving logic base.

"It's also important to have a high tolerance for stress from driving, people contact, and deadlines.

"Salaries vary in each part of the country. In the Southeast, where I work, a tech with experience can expect to negotiate with the company. The starting rate is about $350 per week, or $7.50 to $8.00 per hour. The high end is $600 per week and $12 to $13 per hour. Whether you are paid hourly or earn a salary depends on the company you work for.

"If you like to work with your hands, see daylight, and learn new things every day, this is the job for you."

# INTERVIEW
## Robert Holland
## Owner of a Vending Machine Sales and Repair Business

Robert Holland has worked many years in the vending machine industry and since 1983 has owned his own company, Nothing New, in Phoenix, Arizona, repairing and selling used vending machine equipment.

## How Robert Holland Got Started

"I was attracted to vending from a young age. I was always wondering what made those machines tick. How does a jukebox know where to go to pick up the right record? That sort of thing.

"Every time the vending machine man came into the local pizza parlor, I would be looking over his shoulder to see all I could. He must have thought I was learning something, because he gave me my first job in vending. I filled cigarette machines at first. Later I was shown how to put the new records on the jukebox, and finally I started trying to fix old machines we had in our shop. They were not as hard to figure out as I first thought. In those early years it was much different than today. We didn't just change computer boards; we had to track down the problem and fix it on location. This was back in 1965 in western New York with a family vending business. My duties were filling cigarette machines and collecting all monies from cigarette machines, jukeboxes, pinball machines, and pool tables at various accounts, mostly bars and restaurants.

"When I moved to Arizona in 1981, I found myself beginning a new career in vending—pop, candy, cold food, snacks, the food-service end of vending.

"I started as a street mechanic. That meant that every day I would leave my house and go directly to a service call. Maybe it would be a coin jam, a leaky water valve in a coffee machine, a bad motor in a snack machine, an unlocked door in any machine, and on and on with an endless host of problems that pop up in our business.

"I would also perform preventive maintenance when I wasn't swamped with calls. No one is out there to supervise you, so you are your own boss in the field. I made about $35,000 to $40,000 a year as a repairman, but some national companies, such as Coke and Pepsi, pay higher wages."

## What the Job Is Really Like

"In 1983 I came to own my own company in Phoenix, Arizona, called Nothing New. I buy and sell only used vending equipment that I purchase from vendors that go out of business, retire, or just don't have the space to store excess equipment. I purchase these machines as is, then I totally refurbish them. I then offer them for sale to other vendors.

"I sell my equipment all over the southwestern United States and Mexico. I sell mainly can soda and snack machines, but I also sell dollar-bill changers, food machines, coffee machines, pool tables, video games, and jukeboxes. I have six employees.

"These days I have had to stay up with the endless amount of kits that are coming out to update older equipment so it can still be used. An example of this would be to put a dollar acceptor kit in an older can soda machine so it will accept dollar bills as well as change.

"Also, today computer boards are used extensively in new vending equipment. You don't need to be a computer expert, but you do need to learn the symptoms, indicating when to replace the computer board and when to look for a different problem.

"What I like most about this business is that each employee is pretty much his own boss. If you do your job properly, your time is your own. You can take your breaks whenever you want. Vending isn't on a firm schedule by any means.

"I like having all my weekends to myself. I also enjoy state holidays. A lot of buildings are closed for the day, making my workload for that day pretty easy.

"I have to be honest—there is not much about this business that I don't like. I love what I do, and I have made a very good living in the vending industry—more than $100,000 a year."

## Expert Advice

"My advice to anyone wanting to get started in the vending industry is to look into the companies in your area. Find out which ones are the most reputable, and see if they are looking for any help, even part time. (You can find the companies in the Yellow Pages.) Most prefer you have no experience because they want to train you to their way of running their service department. If you start part time, this will give you an inside look at what makes a good vending company work. Also, this will give you time to find out if this is the line of work you might enjoy. I think smaller operators are preferable to work for;

you only have one person to answer to. You get to know each other on a one-on-one basis. With a larger company, the owner may rarely talk to you and may not realize what kind of job you are doing for him.

"When you start to work for any company, you are low man on the totem pole, so you may have to work some weekends on call. You may also be required to take night calls one week a month or so, depending on the size of the company you work for. Otherwise, it's forty hours a week, five days.

"You will meet a variety of people—mostly really nice people who are very understanding of the problems incurred in vending. If you respond in a timely manner and act in a respectful manner, they will treat you well, but always keep in mind that we are feeding hungry people and we must keep this equipment running day and night.

"I believe this industry offers a way to make a very good living without the need of several years of college. I would recommend this field for most young people who have not found their goals and are undecided as to their future. It will always be a job with openings anywhere in the country at any time."

# FOR MORE INFORMATION

Further information on job opportunities in this field can be obtained from local vending machine firms and local offices of the state employment service.

For general information on vending machine repair, write to:

National Automatic Merchandising Association
20 N. Wacker Drive, Suite 3500
Chicago, IL 60606-3102

American Vending Sales, Inc.
750 Morse Avenue
Elk Grove Village, IL 60007

# About the Author

A full-time writer of career books, Blythe Camenson's main concern is helping job seekers make educated choices. She firmly believes that with enough information, readers can find long-term, satisfying careers. To that end, she researches traditional as well as unusual occupations, talking to a variety of professionals about what their jobs are really like. In all of her books, she includes first-hand accounts from people who can reveal what to expect in each occupation, the upsides as well as the down.

Camenson's interests range from history and photography to writing novels. She is also director of Fiction Writer's Connection, a membership organization providing support to new and published writers.

Camenson was educated in Boston, earning her B.A. in English and psychology from the University of Massachusetts and her M.Ed. in counseling from Northeastern University.

In addition to *On the Job: Real People Working in Mechanics, Installation, and Repair*, Blythe Camenson has written more than four dozen books for NTC/Contemporary Publishing Group.

YA
331.702

M464C

Camenson, Blythe.
On the job.

DEMCO